Global Challenges in Water Governance

Series Editor
Jeremy J. Schmidt
Carleton University
Ottawa, ON, Canada

After a century of massive human interventions into the hydrological cycle, governing water is a critical global concern in the new millennium. Growing evidence that human impacts on the planet are shaping global and local hydrology is challenging long-held assumptions regarding resource management, development, and sustainability. Global Challenges in Water Governance introduces and examines physical, social, and ethical factors that affect how relationships to water amongst humans, social institutions, other species, and Earth systems are governed.

Each volume in the series tackles issues of critical importance to water governance—from relationships of science to policy, to water politics and human rights, to ecological concerns—in order to clarify what is at stake and to organize the complex contexts in which decisions are made. Broadly interdisciplinary, the series provides fresh, accessible insights across the sciences, social sciences, and humanities from established academics and talented young scholars. Individual books are ideal for educators, as policy primers for governmental and non-governmental sectors, and for researchers whose work is directly or incidentally connected to water issues.

More information about this series at
http://www.palgrave.com/gp/series/15055

Kevin Grecksch

Drought and Water Scarcity in the UK

Social Science Perspectives on Governance, Knowledge and Outreach

palgrave
macmillan

Kevin Grecksch
Centre for Socio-Legal Studies
University of Oxford
Oxford, UK

Global Challenges in Water Governance
ISBN 978-3-030-65577-8   ISBN 978-3-030-65578-5   (eBook)
https://doi.org/10.1007/978-3-030-65578-5

This Palgrave Macmillan imprint is published by the registered company Springer Nature Switzerland AG.
The registered company address is: Gewerbestrasse 11, 6330 Cham, Switzerland

Acknowledgements

This book is the result of more than four years of drought and water scarcity research in the UK between 2015 and 2019. The research presented in Chaps. 2 and 4 was part of the MaRIUS project (Managing the Risks, Impacts and Uncertainties of drought and water Scarcity) funded by the UK Natural Environment Research Council (NE/L010364/1). The research presented in Chap. 5 was part of the ENDOWS project (ENgaging diverse stakeholders and publics with outputs from the UK DrOught and Water Scarcity programme) funded by the UK Natural and Environmental Research Council (NE/L01016X/1). Both projects were part of the UK Droughts & Water Scarcity research programme (www.aboutdrought. info). Chapter 3 was funded by the Oxford University John Fell Fund.

I also wish to thank the Centre for Socio-Legal Studies at the University of Oxford for providing an intellectual home and institutional support. I would like to express my deepest gratitude to Bettina Lange, my colleague and mentor at the Centre for Socio-Legal Studies. Without her support, ideas and numerous discussions about the governance and regulation of drought and water scarcity in the UK, this book would not have materialised. I owe similar gratitude to Bernd Siebenhüner for laying the foundations. In addition, I would like to thank my colleagues in the MaRIUS and ENDOWS projects. For Chap. 3, I wish to thank Zita Stefán for her research assistance, especially researching literature and transcribing the interviews. Chapter 5 partially refers to a primer on water efficiency in the public sector and I wish to thank Jessica Holzhausen for her great work on it. For the possibility to teach the issues discussed in this book and for inspiration I would like to thank Stuart Downward.

Special thanks, as always, go out to my parents and l.a.

Kevin Grecksch
Oxford, UK

CONTENTS

ABBREVIATIONS

CCW	Consumer Council for Water
CEH	UK Centre for Ecology & Hydrology
Defra	Department for Environment, Food & Rural Affairs
DP	Drought Plan
DWI	Drinking Water Inspectorate
EA	Environment Agency
EFI	Environmental Flow Indicator
EIA	Environmental Impact Assessment
ENDOWS	ENgaging diverse stakeholders and publics with outputs from the UK DrOught and Water Scarcity programme
EU	European Union
FDF	Food and Drink Federation
HRA	Habitats Regulation Assessment
HTA	Horticultural Trade Association
MaRIUS	Managing the Risks, Impacts and Uncertainties of drought and water Scarcity
NRW	Natural Resources Wales
Ofwat	Office of Water Services—the Water Services Regulation Authority
RSA	Restoring Sustainable Abstraction
SEA	Strategic Environmental Assessments
SEPA	Scottish Environment Protection Agency
SWA	Scotch Whisky Association
TUB	Temporary Use Ban
UK	United Kingdom

UKWIR	UK Water Industry Research
WEF	Water-Energy-Food nexus
WFD	Water Framework Directive
WRMP	Water Resources Management Plan

LIST OF FIGURES

LIST OF TABLES

CHAPTER 1

Introduction

Abstract This chapter explains the rationale of the book, summarises a chapter outline and introduces some of the main concepts used.

Keywords Drought • Water Scarcity • UK • Knowledge • Management • Public engagement

Sir James Bevan, chief executive of England's Environment Agency (EA), underlined in what is now known as 'Jaws of death' speech that unless action is taken to change things, England will not have enough water to supply its needs (Bevan 2019). This book therefore presents a social science perspective on drought and water scarcity in the United Kingdom (UK). The book aims to satisfy several needs. Drought is a recurring feature of the UK climate (Marsh et al. 2007), but it is hardly given attention, and usually flooding receives far greater attention in research but especially in the media. In the last 20 years, the UK underwent drought periods between 2003, 2004–2006 and 2010–2012 (Met Office 2012, 2013, 2016), and the Environment Agency declared an environmental drought in 2019 for Hertfordshire and parts of north London (Environment Agency 2019). In the second half of the twentieth century, the Yorkshire drought of 1995/1996 (Haughton 1998) and most important the drought of 1976, the 'big one' one could say. In all my conversations and

© The Author(s), under exclusive license to Springer Nature Switzerland AG 2021
K. Grecksch, *Drought and Water Scarcity in the UK*, Global Challenges in Water Governance,
https://doi.org/10.1007/978-3-030-65578-5_1

interviews with stakeholders, even if they were born after 1976, this is the drought they remember, and in private conversations it is usually those born before 1976 who remember it. The UK Climate Change Risk Assessment (Committee on Climate Change Risk Assessment 2016) attributes a 'medium magnitude now' but a 'high magnitude in future' for the 'risk of water shortages in the public water supply, and for agriculture, energy generation and industry, with impacts on freshwater ecology'. The overall assessment is that more action is needed in this area.

Garrick et al. (2020) highlight three distinct phases in policy approaches to water scarcity and water quality. In the first phase, the emphasis was on big infrastructure projects such as dams. To some extent this phase is not really over especially in the global south, where big engineering projects such as the Renaissance Dam in Ethiopia or dams on the Mekong River in southeast Asia (Stone 2016) are being planned or built. The second phase focused on treating water as an economic good that could be subject to regulation, prices and the markets. To some extent England and Wales with their full privatisation model can be seen as prime example of this phase, and Helm (2020) discusses whether it was a success or failure. The third phase is described by the authors as a new, beginning phase that will be defined by 'the overdue appreciation that water challenges are inherently multifaceted' (Garrick et al. 2020, p. 3). It is this multifaceted nature of water challenges or water governance that requires a multifaceted, that is multi- if not interdisciplinary approach to drought and water scarcity. However, drought research in the UK, but also beyond, is dominated by the natural sciences, and a great emphasis is on modelling and a tendency to favour supply side measures (see Chap. 2). This of great importance and certainly contributes to making the UK more drought resilient. However, drought and water scarcity also affect society. For example, Temporary Use Bans (TUBs), commonly known as 'hosepipe bans', limit the water one can use and a drought may have large knock-on effects on production processes where water is key (see Chap. 3). And when in drought, it needs to be communicated but ideally issues of drought and water scarcity should be continuously discussed in order to be better prepared.

The book adds to the discussion and argues for a stronger collaboration between natural sciences and social sciences but also the humanities. All chapters in this book argue for a stronger collaboration between stakeholders and policy sectors and the same should happen in academia. The contribution of this book in this regard is to provide a social science perspective on the governance of drought and water scarcity, the generation of drought and water scarcity knowledge, and a perspective on engaging the public with drought and water scarcity (research). The hope is that these individual contributions and the book as a whole stimulate and enrich the necessary discussion about drought and water scarcity.

Drought and water scarcity management in the UK are very specialised and take place in a narrow drought governance space (Lange and Cook 2015). The privatisation of water supply in England and Wales and the relatively low number of water supply companies (~25) are, for instance, an interesting fact. In comparison, Germany has over 6000 water suppliers, approximately 60% of them in public hands and 40% in private hands (Arnold and Pieper 2014). Then, there is the network of seven regulatory bodies regulating the water industry in the UK. Guided by the ministry, the Department for Environment, Food & Rural Affairs (Defra), we find the Environment Agency (EA) for England, Natural Resources Wales (NRW), the Scottish Environment Protection Agency (SEPA), the Northern Ireland Environment Agency (NIEA), Ofwat (Water Services Regulation Authority)—the economic regulator of the water industry in England and Wales, the Drinking Water Inspectorate (DWI) and the Consumer Council for Water (CCW). In addition there are consultancies, who deliver expertise for water companies, and other stakeholders (Cook 2017). Drought and water scarcity are only one aspect of UK water policy. Other aspects are, of course, flooding, water quality and more towards the political end of the spectrum the question of water prices, leakage reduction and the opening up of competition in the water sector similar to the energy sector as it is already happening for business customers or renationalisation, the latter recently fuelled by the Labour party in its election campaigns 2017 and 2019. These are all important aspects, and again, the plead is to look at them integrally. For example, one could ask whether the ownership structure of English and Welsh water companies matters to people and how they value water. Would they value and save more water if it was in public hands? Helm (2020), for example, denies that ownership matters. The public outrage about the salaries and bonuses of water company CEOs speak a different language though (Partington 2018). However, these aspects are not the focus of this book, but the results contribute to the overall discussion about water policy in the UK.

Hence, this book focuses on the governance of drought and water scarcity in the UK, which is of interest for water resources managers, regulators, researchers and students. Although the focus of the book is on the UK, it contains vital results and recommendations for drought and water scarcity management that could be relevant for other countries suffering from drought and water scarcity. This relates, for example, to what and how water suppliers choose and employ drought and water scarcity options. Are the options chosen innovative, do they factor in the human dimension of water governance or are they just following the minimum guidelines provided by a regulatory authority? In return regulators could also up the game by animating water suppliers to be more innovative and inclusive when it comes to drought and water scarcity management options (Chap. 2). Another relevant lesson is which stakeholders are involved in drought and water scarcity governance. This is important for the debate on public participation and the participation of, for example, business and industry (Chap. 3). Participation and access to decision-making and knowledge also shape the relationships of the different actors involved in drought and water scarcity management. This relationship is often shaped by power relationships and who provides knowledge such as monitoring data (see Chap. 4). For instance, one knowledge provider that has often been neglected in previous years and decades is local (expert) knowledge (Chap. 4). By addressing these aspects, this book informs the wider debate on water governance beyond the UK context.

The conceptual focus of this book is (1) on power relationships between drought and water scarcity stakeholders and its implications for the management of both, (2) the creation and management of drought and water scarcity related knowledge and (3) outreach and public engagement with drought and water scarcity research. The different chapters present and discuss research with and about a variety of stakeholders involved (or not involved) in current drought and water scarcity management. Hence the book presents a narrative of how the different stakeholders in drought and water scarcity management manage drought and water scarcity, how they generate and manage knowledge and why some stakeholders are not (yet) involved or only involved in an ad hoc basis. It argues for a more integrated management of water scarcity and drought that includes a strong focus on communication not just during a drought but before and after. A third focus is on public engagement with drought and water scarcity research. Public engagement is two-way process of engaging with different publics about research. This could be helpful in starting the much-needed public discussion about drought and water scarcity and for regaining peoples' value in water.

This book is therefore structured in three parts—governance, knowledge and outreach, plus a brief conclusion. It is the result of researching drought and water scarcity issues in the UK for more than four years between 2015 and 2019. This book answers the following key questions:

- What is the current status of drought and water scarcity management in the UK?
- What is the role of businesses in drought and water scarcity management in the UK?
- How does drought-related knowledge shape power relationships between stakeholders in drought and water scarcity management?
- How to disseminate research results and start a conversation about drought and water scarcity with stakeholders and the public?

In order to answer these questions, this book offers a detailed analysis and critique of all Water Resources Management Plans (WRMP) produced by English and Welsh water supply companies for the period 2014–2019. WRMPs are strategic documents and a statutory requirement. WRMPs look at supply and demand balances over 25 years and lay out plans how to deliver secure public water supplies (Chap. 2). The effect on businesses and production processes is another focus of the book, and the relevant chapter identifies core issues for businesses and industries regarding drought and water scarcity (Chap. 3). The knowledge and especially how knowledge about drought and water scarcity is generated and by whom and for what purpose is another focus of this book and is discussed in Chap. 4. Lastly, there is a need to engage people in the UK about water issues or in other words to start a conversation about drought at places where drought and water scarcity happen but also beyond. In this regard this book offers a perspective on how to communicate and engage with drought research. The author has organised two drought-related walks to discuss and disseminate research results with stakeholders who could act as intermediaries between research and the general public (Chap. 5). Each chapter (2, 3, 4 and 5) can be read on its own and without the necessity to read the whole book. This allows readers who are only interested in specific chapters to fully comprehend them without having to read any of the other chapters. Of course, my hope is that readers are interested in the whole narrative I develop in the book.

Although each chapter introduces the key concepts and its uses, some key concepts that are frequently used in all chapters are defined in the following. *Drought* shares the dilemma with many other concepts and phenomena as there is no consensus over the definition of drought. Lloyd-Hughes (2013) describes it as a 'deficit of water relative to normal conditions', and the UK Environment Agency (Environment Agency 2015) differentiates between an environmental drought, an agricultural drought and a water supply drought. An environmental drought happens when a shortage of rainfall is having a detrimental impact on the environment resulting in reduced river flows, exceptionally low groundwater levels and insufficient moisture within soils. An agricultural drought happens when there is not enough rainfall and moisture in soils to support crop production or farming practices such as spray irrigation. A water supply drought happens when a shortage of rainfall is causing water companies concern about supplies for their customers (Environment Agency 2015, p. 6). What is important though is to emphasise that drought is not just a natural event of limited duration but also a socially constructed event as it can be a result of social factors such as agriculture, housing and transport policies (Lange and Cook 2015). Van Loon et al. (2016) explicitly factor in human processes in drought definitions, an issue that so far has been neglected, according to the authors. Droughts are process specific, local in space, local in time of the year and predicated on the existence of the climatological norm of a process-specific reservoir term (Lloyd-Hughes 2013).

Water scarcity is equally difficult to define. Rijsberman (2006, p. 6) introduces the issue well:

> When an individual does not have access to safe and affordable water to satisfy her or his needs for drinking, washing or their livelihoods we call that person water insecure. When a large number of people in an area are water insecure for a significant period of time, then we can call that area water scarce. It is important to note, however, that there is no commonly accepted definition of water scarcity. Whether an area qualifies as "water scarce" depends on, for instance: (a) how people's needs are defined—and whether the needs of the environment, the water for nature, are taken into account in that definition; (b) what fraction of the resource is made available, or could be made available, to satisfy these needs; (c) the temporal and spatial scales used to define scarcity.

Falkenmark et al. (2007) state that water scarcity does not only result from a physical lack of water but that it is often a sign of difficulties in mobilising more of the freshwater resources available. Among those difficulties are cost, infrastructure-related challenges and the size of the population. The authors also differentiate between blue water scarcity (water from rivers and aquifers) and green water scarcity (water in the soil for crop production). Similar but simpler, van Loon and van Lanen (2013), for example, define water scarcity as the result of long-term unsustainable use of water resources, which water managers can influence. Walker (2014) stresses that water scarcity is hence human induced and subject to the socio-political and economic context. Bakker (2000) puts the natural, social and discursive elements of water scarcity and puts them into the context of water privatisation in England and Wales. In her example of the Yorkshire drought of 1995, she analyses the drought as the production of scarcity in nature, thereby underlining the influence of socio-economic and other factors, in this case the privatisation of water supply in England and Wales.

Governance, again a term difficult to define, is defined for the purpose of this book as steering mechanisms and new modes of coordination, regulation, cooperation and management across multiple levels that include various interdependent actors from politics, economy and civil society aiming at making binding political decisions based on negotiations (Grecksch 2014).

Lastly, a note on the use of 'UK' and 'England & Wales' especially with regard to water management: in common parlance, both terms, but especially 'UK' and 'England', are often used interchangeably, often, but not always, without intention. However, it is important to make the distinction clear. The UK (United Kingdom) consists of four nations: England, Wales, Scotland and Northern Ireland. With the exception of England, the other three have devolved governments with varying powers. With regard to water management, Defra, the UK Ministry for Environment, Food & Rural Affairs, oversees legislation for all four constituent parts of the UK. England and Wales are much alike. Ofwat, the economic regulator for the water industry, for example, is responsible for both nations, but the environmental regulator for England is the Environment Agency, while for Wales it is Natural Resources Wales. Water companies in England and Wales are private, while in Scotland and Northern Ireland, water supply remains in public hands. This book has a strong focus on England and Wales as this was where most of the research took place, but Chap. 3 for example also refers to Scotland. I have tried my best to use the correct term in each case, but I apologise if I did not succeed in every case.

References

Arnold, M., & Pieper, T. (2014). Gesellschaftliche Verantwortung von Wasserwirtschaftsunternehmen. In U. Schrader & V. Muster (Eds.), *Gesellschaftliche Verantwortung von Unternehmen. Wege zu mehr Glaubwürdigkeit und Sichtbarkeit* (pp. 149–177). Metropolis: Marburg.

Bakker, K. J. (2000). Privatizing Water, Producing Scarcity: The Yorkshire Drought of 1995. *Economic Geography, 76,* 4–27. https://doi.org/10.2307/144538.

Bevan, J. (2019). *Escaping the Jaws of Death: Ensuring Enough Water in 2050.* In GOV.UK. Retrieved November 12, 2020, from https://www.gov.uk/government/speeches/escaping-the-jaws-of-death-ensuring-enough-water-in-2050.

Committee on Climate Change Risk Assessment. (2016). *UK Climate Change Risk Assessment 2017.* Synthesis Report: Priorities for the Next Five Years. London

Cook, C. (2017). *Drought Planning in England: A Primer.* Oxford: Environmental Change Institute.

Environment Agency. (2015). *Drought Response: Our Framework for England.* Bristol.

Environment Agency. (2019). *Environmental Drought in Hertfordshire and North London—Creating a Better Place.* Retrieved October 7, 2019, from https://environmentagency.blog.gov.uk/2019/10/01/environmental-drought-in-hertfordshire-and-north-london/.

Falkenmark, M., Berntell, A., Jägerskog, A., et al. (2007). *On the Verge of a New Water Scarcity—A Call for Good Governance and Human Ingenuity.* Stockholm: SIWI.

Garrick, D. E., Hanemann, M., & Hepburn, C. (2020). Rethinking the Economics of Water: An Assessment. *Oxford Review of Economic Policy, 36,* 1–23. https://doi.org/10.1093/oxrep/grz035.

Grecksch, K. (2014). *Adaptive Water Governance. Conclusions and Implications Regarding Adaptive Governance and Property Rights.* University of Oldenburg.

Haughton, G. (1998). Private Profits. Public Drought: The Creation of a Crisis in Water Management for West Yorkshire. *Transactions of the Institute of British Geographers, 23,* 419–433.

Helm, D. (2020). Thirty Years after Water Privatization—Is the English Model the Envy of the World? *Oxford Review of Economic Policy, 36,* 69–85. https://doi.org/10.1093/oxrep/grz031.

Lange, B., & Cook, C. (2015). Mapping a Developing Governance Space: Managing Drought in the UK. *Current Legal Problems, 68,* 1–38. https://doi.org/10.1093/clp/cuv014.

Lloyd-Hughes, B. (2013). The Impracticality of a Universal Drought Definition. *Theoretical and Applied Climatology, 117*, 607–611. https://doi.org/10.1007/s00704-013-1025-7.

van Loon, A. F., & van Lanen, H. A. J. (2013). Making the Distinction between Water Scarcity and Drought Using an Observation-Modeling Framework. *Water Resources Research, 49*, 1483–1502. https://doi.org/10.1002/wrcr.20147.

van Loon, A. F., Gleeson, T., Clark, J., et al. (2016). Drought in the Anthropocene. *Nature Geoscience*. Retrieved February 2, 2018, from https://www.nature.com/articles/ngeo2646.

Marsh, T., Cole, G., & Wilby, R. (2007). Major Droughts in England and Wales, 1800–2006. *Weather, 62*, 87–93. https://doi.org/10.1002/wea.67.

Met Office. (2012). Dry Weather during 2003. In *Met Office*. Retrieved July 28, 2017, from http://www.metoffice.gov.uk/climate/uk/interesting/2003dryspell.html.

Met Office. (2013). England and Wales Drought 2010 to 2012. In *Met Office*. Retrieved July 28, 2017, from http://www.metoffice.gov.uk/climate/uk/interesting/2012-drought.

Met Office. (2016). Dry Spell 2004/6. In *Met Office*. Retrieved July 28, 2017, from http://www.metoffice.gov.uk/climate/uk/interesting/2004_2005dryspell.

Partington, R. (2018). Water Company Bosses' Bonuses Have Damaged Trust, Says Watchdog. *The Guardian*.

Rijsberman, F. R. (2006). Water Scarcity: Fact or Fiction? *Agricultural Water Management, 80*, 5–22. https://doi.org/10.1016/j.agwat.2005.07.001.

Stone, R. (2016). Dam-Building Threatens Mekong Fisheries. *Science, 354*, 1084–1085. https://doi.org/10.1126/science.354.6316.1084.

Walker, G. (2014). Water Scarcity in England and Wales as a Failure of (meta) Governance. *Water Alternatives, 7*, 388–413.

PART I

Governance

CHAPTER 2

UK Drought and Water Scarcity Management Options

Abstract Droughts are a recurring feature of the UK climate. This chapter asks what drought and water scarcity management options are currently applied in England and Wales and contrasts it with options as identified in the literature. The research presented in this chapter will also deviate from the standard differentiation of supply versus demand options and present a new classification focusing on different stakeholder groups or overarching water governance issues. This new typology helps identifying weaknesses in current drought and water scarcity management. The literature review shows a tendency towards proactive measures that focus on cross-sectoral collaboration, abstractor groups and valuing water. In contrast, the results for England and Wales, based on an analysis of water companies' Water Resources Management Plans, show that only a limited number of the available options are applied and that the currently applied options focus on restricting water use in times of drought but focus less on preventing droughts in a larger context of water governance. Thus, to tackle future challenges such as climate change and population growth, water companies and regulators should embrace the introduction of more proactive drought management options.

Keywords Drought • Water scarcity • UK

K. Grecksch, *Drought and Water Scarcity in the UK*, Global Challenges in Water Governance,
https://doi.org/10.1007/978-3-030-65578-5_2

Introduction

Successful management of drought and water scarcity requires the availability of a broad array of management options. Drought and water scarcity management options ensure that sufficient water supply is available during a drought or water scarce situation. This chapter argues that England and Wales lack this broad array of options, but instead English and Welsh water companies stick to the regulatory framework of prescribed options and measures. Speight (2015) judges about the UK water sector: 'The water industry is notoriously slow to implement change, often embracing tradition and tried-and-true methods for achieving their goals.' In her comparison between the US and the UK water sector, Speight concludes that:

> [B]ased on the availability of capital, the UK water companies should be better positioned to implement innovation than publicly funded US utilities. Yet the UK companies need a regulatory driver to justify innovation expenditures within their short payback periods. Ofwat is uniquely positioned to increase spending on innovation and infrastructure replacement, both of which will soon be needed to meet the challenges of increased water demand, high public expectations about service and water quality, and energy efficiency. (Speight 2015, p. 311)

In the context of this research, an innovative drought and water scarcity management option is defined as an option that is not part of the current regulatory framework but as an option that addresses and reflects current trends in water governance such as the multiple uses of water, multi-level governance aspects and a cross-sectoral view on water issues (Gupta et al. 2013; van Loon et al. 2016).

The UK Climate Change Risk Assessment 2017 attributes a 'medium magnitude now' but a 'high magnitude in future' for the 'risk of water shortages in the public water supply, and for agriculture, energy generation and industry, with impacts on freshwater ecology' (Committee on Climate Change Risk Assessment 2016). The overall assessment is that more action is needed in this area (ibid.). 'Drought is a recurring feature of the UK climate' (Marsh et al. 2007). The last drought event was between 2010 and 2012 (Met Office 2013), before that 2004–2006 (Met Office 2016) and 2003 (Met Office 2012). Other major drought events occurred in 1995/1996 and 1976 (Marsh et al. 2007). There is no consensus over the definition of drought. Lloyd-Hughes (2013) describes it as a 'deficit of water relative to normal conditions', and the UK Environment Agency (EA; Environment Agency 2015a) differentiates between an environmental drought, an agricultural drought and a water supply drought. What is important though is to emphasise that drought is not just a natural event of limited duration but also a socially constructed event as it can be a result of social factors such as agriculture, housing and transport policies (Lange and Cook 2015). Van Loon et al. (2016) explicitly factor in human processes in drought definitions, an issue that so far has been neglected, according to the authors. Droughts are process specific, local in space, local in time of the year and predicated on the existence of the climatological norm of a process-specific reservoir term (Lloyd-Hughes 2013). Water scarcity is as difficult to define as drought (see Chap. 1; Rijsberman 2006). A very brief definition of water scarcity is that it is the result of long-term unsustainable use of water resources, which water managers can influence (Van Loon and Van Lanen 2013). Falkenmark et al. (2007) state that water scarcity does not only result from a physical lack of water, but that it is often a sign of difficulties such as cost, infrastructure or population size, in mobilising more of the freshwater resources available. It is hence human induced and subject to the socio-political and economic context (Walker 2014). Bakker (2000), for example, puts the socio-economic factors into the context of water privatisation on England and Wales using the 1995 Yorkshire drought as an example of drought as the production of scarcity in nature.

The management of drought and water scarcity in the UK takes place in a governance space (Lange and Cook 2015) that includes the major actors in UK water governance. In the following, however, this chapter will focus on England and Wales, where the major actors are the Department for Environment, Food & Rural Affairs (Defra), the Environment Agency (EA), Natural Resources Wales (NRW), Natural England, the Water Regulation Services Regulation Authority (Ofwat), private water companies, the Drinking Water Inspectorate (DWI) and the Consumer Council for Water (CCW). All actors operate within a legal framework that is shaped by legislation and regulations such as, for example, the European Union Water Framework Directive (EU-WFD) (EC 2000), the Water Act 2014 (Water Act 2014), the Water Industry Act 1991 (Water Industry Act 1991), the European Union Habitats Directive (EEC 1992) or the EA's Drought Planning Guideline (Defra and Environment Agency 2015; Environment Agency 2015b). In addition, further actors such as the National Farmers Union, the Rivers Trust, local councils and the UK

Irrigation Association have a stake in drought management. A full account of the drought governance space is provided by Lange and Cook (2015).

In addition to droughts being a recurring feature of the UK climate, population growth and increasing water demand are other major pressures the UK water resources face. Thus, drought management options need not only cover aspects of when in a drought but also increase the preparedness for droughts—in other words, proactive measures. Sayers et al. (2017) developed eight golden rules of strategic drought risk management. One rule is to 'implement a portfolio of measures to transition towards a drought resilient society' (ibid.). Robins et al. (2017) would like to see the creation of a more water-literate society that will better enable water managers to shift from reactionary, crisis-driven approaches to long-term, agenda-driven plans in line with agreed strategies. This chapter picks up on these points and will demonstrate that England and Wales have a long way to go with regard to drought and water scarcity management in order to become a drought resilient society and a water-literate society.

This chapter asks the question what drought and water scarcity management options are currently implemented and applied in the English and Welsh context and juxtapose it with drought and water scarcity management options as identified in academic literature, documents such as policy reports or government documents. This chapter will also deviate from the standard differentiation of supply versus demand options in drought and water scarcity management and instead present a novel typology that could potentially help to broaden the array of available options, and it could help to shift attention to areas in need when implementing new options and measures. This research is the first to analyse all English and Welsh water companies regarding their Water Resources Management Plans (WRMPs) and drought and water scarcity management options. Although limited in focus geographically, this research, especially the newly developed classification of drought and water scarcity management options, is potentially useful beyond the English and Welsh case presented here.

Data and Methods

The data presented in this chapter originates from two major sources: first, a non-exhaustive national and international literature and document review of drought and water scarcity management options and second, an analysis of all English and Welsh water companies' Water Resources

Management Plans (WRMPs). The aim of this approach is to contrast currently implemented drought and water scarcity management options in England and Wales with available options identified through a literature and document review in order to get a better picture of where English and Welsh drought and water scarcity management currently stands.

The literature review included not only academic literature but also government documents and grey literature, for instance, research project reports. The literature review focused on western, industrialised countries such as the United States, Australia or European Union countries in order to provide comparability with the UK. Although there are countries that are far more affected by drought than those named above, for example in Africa (WMO 2014), they hugely differ in impacts and measures against drought and water scarcity. The literature and document review was non-systematic. Literature and documents were searched using Web of Science, Scopus and World Wide Web search engines. All literature, documents and research project websites were searched to identify drought and water scarcity management options. Papers and documents were selected on the basis of dealing with drought and water scarcity management options and a snowball search using cross-references but also the author's previous experience in the field. This includes management options and strategy for water efficiency, how to balance supply and demand, leakage reduction and preventions as well as metering among others. Examples of search terms include 'drought management', 'water scarcity management' and 'drought planning'. The search and review was carried out in spring 2016, but more recent papers and documents were added during the analysis phase if found relevant to the research question. All together 50 academic journal articles, documents and reports published between 2000 and 2016 were analysed, and 4 major European research projects on drought and water scarcity and their results were also included. However, even this limited scope provided a wealth of insight on the issue. Table 2.1 presents an overview over the analysed literature and documents.

The second source is an analysis of all current English and Welsh water companies' Water Resources Management Plans (WRMPs). Water companies in England and Wales are important because they 'occupy a central, powerful position in the governance space' (Lange and Cook 2015). Since 1989 all water companies in England are privately owned. Welsh Water, which supplies water to most parts of Wales, is a company that has no shareholders and is run for the benefit of its customers and hence the only exception to the privately owned model. WRMPs are a statutory

Table 2.1   Overview of the analysed literature and documents

Academic literature
Bokal et al. (2014); Farmer (2012); Garrote et al. (2007); Gleick et al. (2011); Gleick
and Heberger (2012); Gómez Gómez and Pérez Blanco (2012); Horne (2016); Ingram
and Malamud-Roam (2013); Kampragou et al. (2011); Kron et al. (2016); Lange and
Cook (2015); Lorenzo-Lacruz et al. (2010); Marsh et al. (2007); Nelson et al. (2008);
Pérez-Urdiales and García-Valiñas (2016); Priscoli and Hiroki (2016); Rossi and
Cancelliere (2013); Stakhiv et al. (2016); Stone (2014); van Loon and van Lanen (2013);
Wilhite (2002); Wilhite et al. (2014); Wilhite et al. (2007); Zetland (2016)
Government documents and reports
UK:
Environment Agency Water Company Drought Plan Guideline (Environment Agency
2015b)
Drought permits and drought orders. Information from the Department of Environment,
Food & Rural Affairs, Welsh Assembly Government and the Environment Agency
(Department of Environment, Food & Rural Affairs, Welsh Assembly Government, and
Environment Agency 2011)
Drought response: our framework for England (Environment Agency 2015a)
Managing Water Abstraction (Environment Agency 2016)
California (United States)
California's Drought: Water Conditions and Strategies to Reduce Impacts. Report to the
Governor on 30 March 2009 (Department of Water Resources and Department of Food
And Agriculture 2009)
California Drought Contingency Plan (State of California, Natural Resources Agency and
California Department of Water Resources 2010)
United States
Effects of Drought on Forests and Rangelands in the United States: A Comprehensive
Science Synthesis (Vose et al. 2016)
Australia
Water Markets in Australia. A Short History (Australian Government. National Water
Commission 2011)
Agricultural Competitiveness White Paper (Australian Government 2017)
Grey literature (reports and policy briefs)
WMO/GWP. National Drought Management Policy Guidelines: A template for action
(World Meteorological Organization [WMO] and global Water partnership [GWP]
2014)
UK
Water resources long-term planning 2015–2065 (Water UK 2016)
Markets, water shares and drought: Lessons from Australia. What can the water industry
in England and Wales learn from Australia's water reform story? (Piure 2014)
Water resources in the Southeast. Progress towards a shared water resources strategy in
the South East of England (Critchley and Marshallsay 2013)

(*continued*)

Table 2.1 (continued)

California
Allocating California's Water: Directions for Reform (PPIC 2015)
Managing California's Water: From Conflict to Reconciliation (Hanak et al. 2011)
Economic Analysis of the 2015 Drought for California Agriculture (Howitt et al. 2015)
Sustainable Water and Environmental Management in the California Bay-Delta
(Committee on Sustainable Water and Environmental Management in the California
Bay-Delta et al. 2012)
Drought research projects
Xerochore (URL: http://www.feem-project.net/xerochore/)
DROP project (URL: https://www.vechtstromen.nl/projecten/projecten/noordwest-
europees/kopie-noordwest/)
WATCH project (URL: http://www.eu-watch.org/)
Drought R&SPI (URL: http://www.eu-drought.org/)

requirement that water companies write every five years. WRMPs look at supply and demand balances over 25 years and lay out plans on how to deliver secure public water supplies. As such they are strategic documents, approved by regulators. Hence, they are an important, credible and valuable source for analysis. Table 2.2 provides an overview over the analysed WRMPs. The analysis searched for drought and water scarcity management options in all WRMPs and considered how water companies ensure sufficient deployable output, strategies on metering, leakage, water efficiency and other key strategies for efficient use or to reduce pollution plus any other noteworthy items that did not fall into these categories. The analysis did consider not only options that are currently implemented and applied but also options which are envisaged for later implementation, since the timeframe of a WRMP is 25 years. WRMPs are strategic documents and were favoured in the analysis over Drought Plans, another statutory requirement. WRMPs are broader in terms of water resources management, outward looking and hence more interesting to answer the research question, whereas Drought Plans are operational plans describing actions necessary to deal with various drought situations. They set out how a water company will continue to meet its duties to supply water during drought periods. However, all water company Drought Plans are based on the Environment Agency's Drought Plan Guideline (cf. Table 2.1), which was part of the analysis.

All literature, documents and WRMPs were analysed using qualitative content analysis (Bryman 2012; Mayring 2008). The analysis of the data produced an understanding of drought and water scarcity management

Table 2.2 Overview of the analysed Water Resources Management Plans in England and Wales

Water company	Document title	Source
Affinity Water	Our Plan for Customers & Communities. Final Water Resources Management Plan, 2015–2020	Affinity Water (2014)
Albion Water	Draft Water Resource Management Plan	Albion Water (2016)
Anglian Water	Water Resources Management Plan 2015	Anglian Water (2014)
Bournemouth Water (Sembcorp)	Water Resources Management Plan Final Water Resources Management Plan—2014 Technical Report	Sembcorp Bournemouth Water (2015)
Bristol Water	Water Resources Management Plan 2014	Bristol Water (2014)
Cambridge Water (South Staffs Water)	Water Resources Management Plan 2014. Cambridge Region. Main report	Cambridge Water (2014)
Cholderton and District Water	Water Resources Management Plan 2014	Cholderton and District Water (2014)
Dee Valley Water	Water Resources Management Plan December 2013	Dee Valley Water (2013)
Dwr Cymru (Welsh Water)	Final Water Resources Management Plan. Technical Report	Welsh Water (2014)
Essex & Suffolk Water (Northumbrian)	Final Water Resources Management Plan 2014	Essex & Suffolk Water (2014)
Hartlepool Water (Anglian Water)	Water Resources Management Plan 2015	Anglian Water (2014)
Northumbrian Water	Final Water Resources Management Plan 2014	Northumbrian Water (2014)
Peel Water networks	Revised Draft Water Resources Management Plan 2013	Peel Water Networks (2013)
Portsmouth Water	Final Water Resources Management Plan 2014	Portsmouth Water (2014)
Severn Trent	Final Water Resources Management Plan 2014	Severn Trent (2014)
South East Water	Water Resources Management Plan	South East Water (2014)
South Staffs Water	Water Resources Management Plan 2014. Main Report	South Staffs Water (2014)
South West Water	Water Resources Management Plan	South West Water (2014)
Southern Water	Water Resources Management Plan 2015–40. Technical Report	Southern Water (2014)

(*continued*)

Table 2.2   (continued)

Water company	Document title	Source
SSE Water	Water Resources Management Plan (England) 2015–2040. SSE Water. Revised Draft Consultation Water Resources Management Plan 2015–2040. SSE Water. Draft Consultation	SSE Water (2014a, b)
Sutton and East Surrey Water	Final Water Resources Management Plan. Main Report	Sutton and East Surrey Water (2014)
Thames Water	Final Water Resources Management Plan 2015–2040	Thames Water (2014)
United Utilities	Final Water Resources Management Plan. March 2015	United Utilities (2015)
Veolia Water Projects	Water Resources Management Plan. Final published report	Veolia Water Projects (2014)
Wessex Water	Final Water Resources Management Plan. Website version	Wessex Water (2014)
Yorkshire Water	Water Resources Management Plan	Yorkshire Water (2014)

options, and it included the identification of key themes and patterns that emerged from reading the literature, documents and WRMPs. Themes are recurring ideas, issues or statements expressed in the data, however, often not directly. Hence, identifying themes can help to uncover further dimensions and facets of, in this case, drought and water scarcity management. The identified themes have, in the next step, been 'translated' into a new typology of drought and water scarcity management options (see Table 2.3).

Based on this, a further aim was to overcome the traditional differentiation between supply and demand side options. It is argued that this dichotomy falls short of successfully identifying and pointing out deficiencies in drought and water scarcity management options, because it is too general. As highlighted in the introduction, drought and water scarcity management options need to include socio-economic and human dimensions, and the newly developed typology can help to bring them better to light than a long list of supply and demand options. This typology does not try to abolish the demand and supply dichotomy but to complement it. Hence, this typology helps identifying where the emphasis in current drought and water scarcity management lies, and it helps pointing out

Table 2.3   Typology of drought and water scarcity management options

Criteria	Description
Current regulatory framework	This category describes options as prescribed by the current legal and regulatory framework. Examples: Drought orders and drought permits
Structural approaches/overarching framework	This category reflects overarching measures, that is measures that go beyond the individual option and instead, for example, aim for networks and collaborations with other sectors. Examples: Collaboration with other policy sectors such as housing or international legislation
Abstractor groups	This addresses options aimed at specific water abstractor groups such as agriculture or large business abstractors (e.g. power sector and paper industry). Examples: Irrigation management and agricultural insurance
Procedural devices	A procedural device refers to options such as water markets
Distribute water	This category includes options that put emphasis on sharing and transporting water. Examples: Bulk water agreements or sharing water
Supply side/water creation	This category contains options aiming to provide more water. Examples: Water recycling, reservoirs or desalination
Metering	Metering is a key option and includes current trends such as smart metering and tariffs
Technology led	This category describes options driven by technology such as deepening a borehole
Land use planning	This category refers to option that takes a view beyond drought and water scarcity management and integrates it into the wider discussion on land use planning. Examples: Restoring wetlands or drought tolerant landscaping
Company led	This describes options implemented at the discretion of water companies such as pressure management
Valuing water/water attributes	This category refers to options that stress societal aspects of drought and water scarcity management. Examples: Water stewardship or water rights

weak points, that is areas that could and should potentially be given more attention in the future. In this regard it could also be viewed as a learning tool for water companies and regulators. Beyond the UK context, the typology has potential to act as toolbox for water companies and water authorities. For example, even if costly options such as desalination are not feasible due to financial constraints, water companies or regulators might opt for other, less costly options to reduce water consumption such as water efficiency campaigns or handing out water saving devices. The limitations of the typology are discussed below.

RESULTS

Water governance is a complex issue and, as mentioned in the introduction, faces several challenges such as climate change, population growth and linked to that rapid urbanisation. This implies major challenges for water supply, wastewater treatment, flooding and drought policies. Robins et al. (2017, p. 41) begin their assessment of water policy in the UK, but the statement could also be applied to other nations, by stating that 'water resources management is integral to economic, ecological and socio-political sustainability, its management is complex and requires coordination across a range of institutions and stakeholder interests.' Although the management of drought and water scarcity is just one aspect, among others, in water governance, the above statement is also true just within the sub-field of drought and water scarcity management. The results from the literature and document review support this view and reveal a broad array of drought and water scarcity management options. There is a clear tendency towards proactive measures that focus on cross-sectoral collaboration such as catchment management, integrating water scarcity into planning or the collaboration of water suppliers with neighbouring policy fields (Farmer 2012; Hanak et al. 2011; Kampragou et al. 2011; Wilhite 2002). Other options pay attention to certain abstractor groups such as farmers and include items such as agricultural insurance or income support (Nelson et al. 2008). Another set of options puts emphasis on the value of water, for instance, water stewardship or the creation of water saving cultures (Farmer 2012). These options are the way forward in enhancing resilience to the increased risk of drought and water scarcity. Figure 2.1 presents the results of the literature and document review and is arranged according to the typology introduced in the previous section. It is a non-exhaustive collection of drought and water scarcity management options, and a brief description of each option based on the literature review is provided in Table 2.4 in the appendix to this chapter.

Against the background of the drought and water scarcity management options identified in the literature and document review, the current range of drought and water scarcity management options in England and Wales is presented in Fig. 2.2. All encircled options are either currently applied or the implementation is planned in the future. The analysis is based, as described under 'Data and methods', on the analysis of WRMPs of all English and Welsh water companies (Table 2.2). The overview does not reflect how many water companies apply a certain method; hence, even if

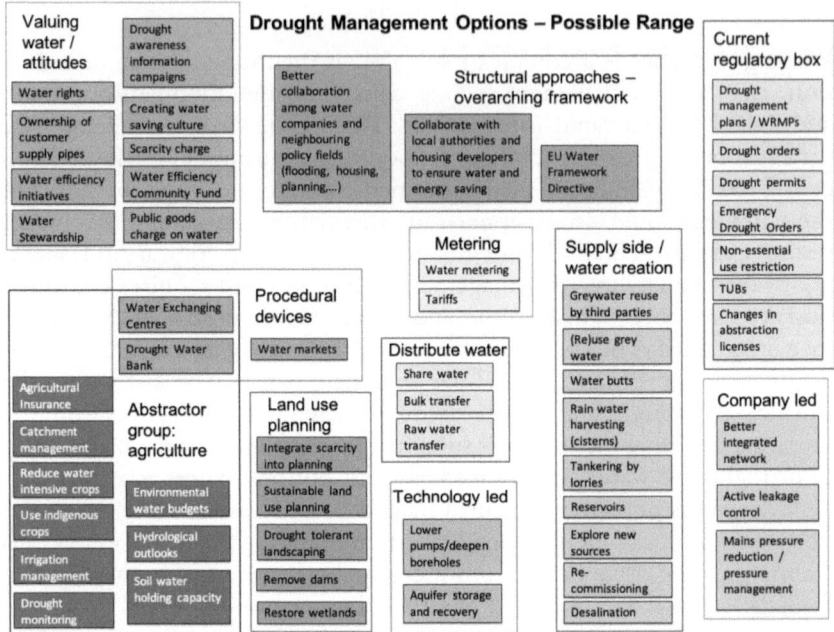

Fig. 2.1 Drought and water scarcity management options based on literature and document review

just one water company has implemented a certain measure, it is reflected in the illustration.

The illustration shows that English and Welsh water suppliers are using only a fraction of the options available given that this is an optimistic overview, that is as mentioned before options that are foreseen for future implementation, but potentially run the risk of being dropped for the next round of WRMPs, are also included. Figure 2.2 highlights a tendency towards the current regulatory box and supply side options. Thus, it can be concluded that currently employed drought management options in England and Wales rely significantly on restricting water use in times of drought and are therefore, with the exception of elements of Drought Plans and WRMPs, potentially too much focused on thinking about water scarcity in the context of actual drought events. Based on the WRMP analysis, all English and Welsh water companies apply a standard set of options that include metering, active leakage control or water efficiency campaigns. Apart from options in the 'current regulatory box', these are the options water companies aim for the most, although with varying degrees. For example, metering strategies are influenced by the Environment Agency's classification of an area as being under 'severe water stress' (Environment Agency and Natural Resources Wales 2013). If an area is classified as 'under severe water stress', metering becomes compulsory. All newly built properties are also fitted with a water meter, and water companies often opt for meter installation when a property changes occupancy or they offer for voluntary metering programmes. Hence, meter penetration differs largely across the country or even within water companies different water resources zones. Affinity Water's Southeast region has a meter penetration of 93%, while its Central region has only 42% meter penetration (Affinity Water 2014). Northumbrian Water currently has a very low meter penetration, 30%, but like other water companies, it aims to increase metering in the next 25 years (Northumbrian Water 2014). Besides the options prescribed by the regulatory bodies, a few water companies either have implemented or are contemplating options that go beyond the required; for example, the 'water efficiency community fund', which provides the installation of water saving devices in public buildings such as schools (Wessex Water 2014), or the concept of the 'scarcity charge' (Southern Water 2014), which would introduce a higher price to be paid for water which is abstracted from areas where there is less water available. Portsmouth Water (2014) and Cambridge Water (2014) highlight the benefits of grey water (re)use. Efforts to collaborate with other sectors such as the housing or energy sector in order to contribute to overall energy savings are also noteworthy (Essex & Suffolk Water 2014).

Although the list of drought and water scarcity options currently applied in England and Wales is far from being incomplete or insufficient, contrasting Figs. 2.1 and 2.2 could leave the impression English and Welsh water companies do miss out on options, especially proactive measures, like creating a water saving culture, collaboration with other sectors or sustainable land use planning. However, it is important to say that all water companies undergo a so-called option appraisal stage when they develop their WRMPs, described in the Environment Agency's Guideline: 'a water company should decide on the best option for its customers (on the basis of cost and what customers would like) and for the environment (local and global). Deciding which option to choose is known as 'option appraisal' and must include both monetary and non-monetary costs and benefits' (Environment Agency and Natural Resources Wales, 2016). Water companies such as South East Water (2014) provide impressive lists of options during this stage, which in itself contains four steps—from an unconstrained list of options to a constrained list, followed by feasible and preferred list of options. In the case of South East Water, this meant 912 unconstrained options, which were already reduced to 320 in the feasible list of options (ibid.). Usually these are not, as in this case 912 distinct options, but often variants of one option. For example, the augmentation of a borehole can be described in numerous small increments, each resulting in a distinct option. A deeper understanding as to why options fall off the list was not focus of this research, but the description of the options appraisal stage by the Environment Agency (see above) might provide a clue, thus being either little willingness to pay by customers or high implementation costs for water companies.

Discussion

The typology for drought and water scarcity management options (Figs. 2.1 and 2.2) does come with limitations. First, it is a non-exhaustive list. The literature and document review was focused and geographically limited as explained above. Also, some options are generic, in the sense that they only provide a name for a category of options. A good example is 'water efficiency campaigns', which can be subdivided into many different items such as websites, blogs, school plays and roadshows (Grecksch and Lange 2019). For simplicity reasons not all options that would fall under one category were presented. Second, it reflects the situation in England and Wales, which is most strongly reflected in the 'current regulatory box' category. In order to be useful in other contexts, it would need to be adapted accordingly. However, for example, in a European Union context, this would be easily achievable as the Water Framework Directive affects every EU member state. Third, not all of the options are mutually exclusive and there are overlaps. For example, 'drought water banks' or 'water exchange centres' are predominantly found in the agricultural sector but are procedural devices that may also be applied in other sectors. Another limitation is that not every option can be applied and implemented by every water company due to, for example, geographical reasons. Desalination might be a viable option for water companies operating at the coast but not for water companies with no access to salt or brackish water. Yet, the idea of the typology is not that every option is adopted, but rather that it is seen as toolbox of available options. One goal of this chapter was to show that the UK water sector limits itself in its selection and application of drought and water scarcity management options and that water companies limit themselves to the regulatory framework and the options prescribed by it. The idea of the developed typology is to demonstrate that other options can be embraced and be taken into account by water companies and regulators. This could support future-proofing the UK water sector against mentioned challenges such as population growth and climate change.

Thus, this novel typology proves effective as it can display weaknesses but also strengths in current drought and water scarcity management in a given jurisdiction. Water companies can use the typology for the same purpose, to identify strengths and weaknesses; hence, it could potentially be used to explore areas for future improvement and innovation.

Drought planning by water companies has become a major feature of water resource planning (Cook 2016). Yet, as the results show, compared to available options for drought and water scarcity management, English and Welsh water companies only apply a fraction of them. They orientate themselves at the regulatory framework, that is the options prescribed by the regulatory bodies. Against the background of the UK Climate Change Risk Assessment 2017, which concluded that more action is needed in the area of public water supply, agriculture, energy and industry (Committee on Climate Change Risk Assessment 2016), it is worth asking the question whether the current set of options will be enough to tackle future water challenges. This addresses especially the result that currently applied options focus on restricting water use in times of drought but focus less on preventing droughts in a larger context of water governance. However, the UK Water Act 2014 introduced a so-called resilience duty to achieve long-term resilience of water and wastewater systems and service provision. This includes the use of 'a range of measures to manage water resources in sustainable ways' (UK Water Act, 2014). Key features of resilience are diversity and redundancy (Folke et al. 2005; Walker et al. 2004). Hence, the availability of a diverse set of options is favourable and also opens up opportunities to address drought and water scarcity issues in a different way. Figure 2.1 can actively contribute to this discussion.

Furthermore, the section on the resilience duty says, 'increase efficiency in the use of water and reduce demand for water so as to reduce pressure on water resources' (UK Water Act, 2014). Demand management is more challenging for water companies than supply side options because they have to engage with the socio-political context (Walker 2014). By focusing more on demand management, water companies could not only broaden their set of options but, by increasing their engagement with the public, also help building a drought-aware society, if not to say water-literate society (Robins et al. 2017), thereby changing the behaviour citizens, and not just water company customers, show towards water.

The majority of UK water companies collaborate among each other through bulk water agreements. So far, only three examples of water company collaboration go beyond this. The Water Resources in the South East Group (WRSE), Water Resources East Group in Anglia and Water Resources North (WReN). All three organisations foster the collaboration between water companies, regulators and other stakeholders in the respective regions, but they lack a wider stakeholder inclusion that could bring fresh perspectives into the groups. Thereby they are neglecting recent trends in water research such as the nexus approach (Green et al. 2017; Gupta et al. 2013) or catchment-based management (Robins et al. 2017). The same holds true for collaborations with other policy sectors. Flooding, agriculture, forestry and housing are just a few of the many policy fields that are highly interconnected with the water sector and could be given more attention by water companies. However, personal communications with representatives from water companies about the results of this research indicate that future drought and water scarcity management options should reflect current trends in water resources management such as more collaboration among water companies, regulators and stakeholders more strongly.

CONCLUSION

This chapter established an overview over the currently applied drought and water scarcity management options in England and Wales and contrasted it with available options based on a literature and document review. The results confirm that English and Welsh water companies adhere to options that are prescribed by the regulatory framework and options that favour the supply side of water resources management. Overall, it can be concluded that drought and water scarcity options in England and Wales are reactive rather than proactive. This chapter also established a new typology for drought and water scarcity management options that differs from the usual supply and demand dichotomy. Instead, the typology introduces a new classification that covers aspects such as options according to different abstractor groups, overarching frameworks or valuing water. It was argued that this novel typology could help identifying weak points in current drought and water scarcity management, thereby opening up the opportunity to introduce new options. If drought is a recurring theme of UK climate, then water companies should more actively embrace new options and should also consider going beyond the 'willingness to pay' and 'cost-benefit analysis' horizon in order to tackle future challenges such as climate change, population growth and changing water demand patterns.

Appendix

Table 2.4 Overview and brief description of drought and water scarcity management options

UK current regulatory box	
Drought orders	A drought order, issued by the Secretary of State (Defra) or the Welsh Minister, authorises increased abstraction from the environment by water companies or any other abstractor in order to meet statutory duties for public water supply. It can also restrict the demand from commercial water users or limit abstraction by a water company or the EA.
Drought permit	A drought permit, issued by the EA, enables to increase supply of water abstracted from the natural environment.
Drought plan	Drought plans cover the range of actions necessary to deal with various drought situations. They set out how a water company will continue to meet its duties to supply water during drought periods with as little recourse as possible to drought permits or drought orders.
Water Resources Management Plan	WRMPs look 25 years ahead and conform to UK legislation and Environment Agency guidelines to ensure companies have sufficient water to supply the public and maintain adequate water in the environment.
Temporary Use Bans/ non-essential use restrictions	Water companies can implement temporary water use restrictions under their own powers. These restrictions are temporary measures that reduce the demand for water and are usually one of the first steps a water company can take to protect its supplies during a drought. The water company does not require any approvals to restrict these uses of water but must run a period of public notice and allow for representations to be made before the restriction comes into force. Examples: watering gardens, cleaning a private motor vehicle

(continued)

Table 2.4 (continued)

Abstraction licences	Abstraction licences provide abstractors with a licence to take a fixed volume of water
Emergency Drought Orders	Emergency Drought Orders may prohibit or limit the use of water for any purposes a water company considers appropriate. Water is supplied by means of standpipes or water tanks.
Abstractor group: agriculture	
Catchment management	A catchment-based approach looks at activities and issues in the catchment as a whole, rather than considering different aspects separately in different locations. It involves bringing people together from different sectors to identify issues and agree priorities for action—and ultimately building local partnerships to put these actions in place.
Soil water holding capacity	One of the main functions of soil is to store moisture and supply it to plants between rainfalls and irrigation. Evaporation from the soil surface, transpiration by plants and deep percolation combine to reduce soil moisture status between water applications. If the water content becomes too low, plants become stressed.
Irrigation management	Irrigation is the artificial exploitation and distribution of water at project level aiming at application of water at field level to agricultural crops in dry areas or in periods of scarce rainfall to assure or improve crop production.
Drought monitoring	Early warning systems supported by data networks, data sharing, forecasts, Standardised Precipitation Index (SPI), and so on.
Agricultural insurance	Agricultural insurance covers yield losses caused by droughts.
Reducing water intensive crops	Reducing water intensive crops means giving preference to crops that use less water in water scarce regions.
Environmental water budgets	Environmental water budgets account for the inputs, outputs and changes in the amount of water by breaking the water cycle down into components.
Indigenous crops	Indigenous crops have their origin in the area they are grown in.
Income support	An income support scheme provides eligible farmers and their partners who are experiencing financial hardship due to a drought with assistance and support to improve their long-term financial situation. Example: Australia.
Interest rate subsidies	Interest rate subsidies provide business support to farms that were viable in the long term, but were in financial difficulties due to a drought event. Example: Australia.

(*continued*)

Table 2.4 (continued)

Hydrological outlooks	Hydrological outlooks are based on observed data and projections, and they present the UK water situation for the next one to three months and beyond.
Procedural devices	
Water exchanging centres	Water exchanging centres offer and demand water use rights in periods of drought. Example: Spain.
Drought water bank	Drought water banks allow selling and buying of water.
Water markets	Water markets provide a more flexible allocation of water. In the Murray-Darling Basin (Australia), the poster child for water markets, water licences used to be tied to land, but water markets allowed to get water you need from someone who already has a licence. However, market rules need to reflect hydrological realities.
Distribute water	
Bulk transfer	Bulk transfer is the transfer of raw or treated water between two parties, for example water companies and areas. Bulk transfers are usually subject to contracts between the two parties.
Share water	Sharing water encompasses a wider concept than bulk transfers. Sharing water can mean voluntary sharing of water resources across areas or sharing and (re)using water for different purposes, that is irrigation and production processes.
Supply side/water creation	
Grey water reuse by third parties (inset appointees)	In this special case third party suppliers, who, for example, supply and treat water for a new housing development, invest in grey water reuse schemes, thereby decreasing future bulk water transfers from the local statutory water supplier.
(Re)use grey water	This process describes the utilisation of treated or untreated water for a variety of purposes. For example, household discharge could be reused for non-potable uses such as watering gardens.
Explore new sources	This includes tapping into aquifers, new river abstraction points, or can go as far as transporting water from geographical distant regions by ship.
Rainwater harvesting	Rainwater harvesting is the accumulation of rainwater for reuse (e.g. irrigation). This includes, for example, cisterns or collection from roofs.
Reservoirs	Reservoirs are artificially created lakes for storing water. Reservoirs are fed by rivers or glaciers and usually provide drinking water and irrigation water. Reservoirs and dams are also used to generate electricity through hydropower.

(*continued*)

Table 2.4 (continued)

Water butts	A water butt or tank is used to collect rainwater runoff usually from rooftops. The collected water can, for example, be used for watering gardens.
Tankering by lorries	This describes the provision of usually drinking water by means of water tanks in emergency situations.
Desalination	Desalination describes the process of removing salt from saline water (sea water and brackish water) either through thermal desalination or through reverse osmosis. Desalination plants are energy intensive, and so far in the UK only Thames Water operates a desalination plant for emergency purposes.
Re-commissioning	Re-commissioning of sources is the process or reactivating previously closed down boreholes or other abstraction points. This could be the case, for example, if a groundwater aquifer has recharged.
Technology led	
Aquifer recharge and aquifer storage and recovery	Aquifer storage and recovery is a process to convey water underground. Aquifer recharge replenishes groundwater stored in aquifers. Aquifer storage and recovery is used to store water and reuse it at a later stage.
Lower pumps/deepen boreholes	This option is applied when the groundwater table drops below the level of the well pump.
Company led	
Mains pressure reduction/ pressure management	Pressure is the force that pushes water through pipes. Water companies apply pressure management to reduce leakage and thereby reduce the loss of water.
Better integrated network	A better integrated network means improved links between water resource zones. This enables water suppliers to distribute water more efficiently and allocated it to where it is needed.
Active leakage control	Active leakage control aims at prompting detection, localisation and repair of pipe burst, thus reducing possible damages to properties, minimise unplanned works and reduce the volume of lost water.
Valuing water/water attributes	
Water efficiency campaigns	Water efficiency campaigns aim at reducing water demand. They usually address private household customers but also aim business customers. Water efficiency campaigns use a variety of media—print, audio-visual and social—or take the form of plays, games or roadshows. Another form of water efficiency campaigns is the provision of water saving devices to customers.

(*continued*)

Table 2.4 (continued)

Water stewardship	Water stewardship describes the use of water that is based on social equality, sustainability, yet economically beneficial approach. It includes a wide variety of stakeholders and is catchment based.
Water efficiency community fund	A water efficiency community fund provides devices and their installation in schools and other not-for-profit social organisations such as hospitals, councils and local services.
Ownership of customer supply pipelines	The ownership of customer supply pipelines enables water companies to target leakage reduction according to their own strategies and timescales.
Water rights	A water right describes the right to use water from a source (e.g. groundwater, river, etc.).
Scarcity charge	A scarcity charge would mean that the price abstractors pay better reflects the environmental impact of water abstraction. If introduced, a higher price would be paid for water, which is abstracted from areas where there is less water available.
Creating water saving culture	A water saving culture promotes water efficiency and tackles issues of reduced supply and increased demand.
Public goods charge on water	A public goods charge for water users funds improvements in water infrastructure and environmental protection and conservation and restoration and research.
Drought awareness (information campaigns)	A drought awareness campaign proactively promotes water efficient behaviour before and during a drought. This can take the form of educational measures or tangible items such as providing buckets for rainwater collecting.
Structural approaches/overarching framework	
EU Water Framework Directive	The EU Water Framework Directive is a comprehensive approach to address qualitative and quantitative issues with regard to water. It addresses chemical issues, promotes public participation and requires river basin management plans. The directive's goal is to achieve a good ecological status of all European Union water bodies.
Collaborate with local authorities and housing associations/developers	Collaboration between water companies and local councils and housing developers can make sure that new and refurbished housing benefits from the latest water saving technologies and appliances.
Regional drought preparedness networks	Regional drought preparedness networks allow sharing information, technologies and tools.

(*continued*)

Table 2.4 (continued)

Better collaboration among water companies and neighbouring policy sectors (flooding, agriculture, forestry, housing, etc.)	A better collaboration can ensure that policies are aligned, and measures in one sector do not contradict measures in another sector. It also raises problem awareness among sectors and its stakeholders.
Metering	
Tariffs	Tariffs are a measure to incentivise efficient water usage. Water tariffs can be volumetric, that is metred, or a flat rate. With regard to consumption, different models exist, for example increasing block tariffs, where tariffs increase with consumption.
Water metering	Charging customers according to use.
Land use planning	
Restore wetlands	The restoration of a wetland describes the return of an ecosystem to a close approximation of its condition prior to disturbance.
Remove dams	Water stored in reservoirs is subject to evapotranspiration, and there are more cost-effective ways to store water available. In addition, the removal of a dam can restore a negatively affected ecosystem.
Drought tolerant landscaping	Drought tolerant landscaping takes into account the ecological characteristics of each space. It includes using efficient irrigation, the use of native plants, use of succulents and strategic plant grouping.
Sustainable land use planning	Sustainable land use planning takes into account regional, social, ecological and economic characteristics. For example, it takes into account population forecasts, and planning is based on preserving the liveability and environmental protection.
Integrate water scarcity in planning	Integrating water scarcity into planning means to acknowledge that water scarcity is part of life in a given area, and thus, water scarcity should not be seen as a problem during emergencies but as a constant condition.

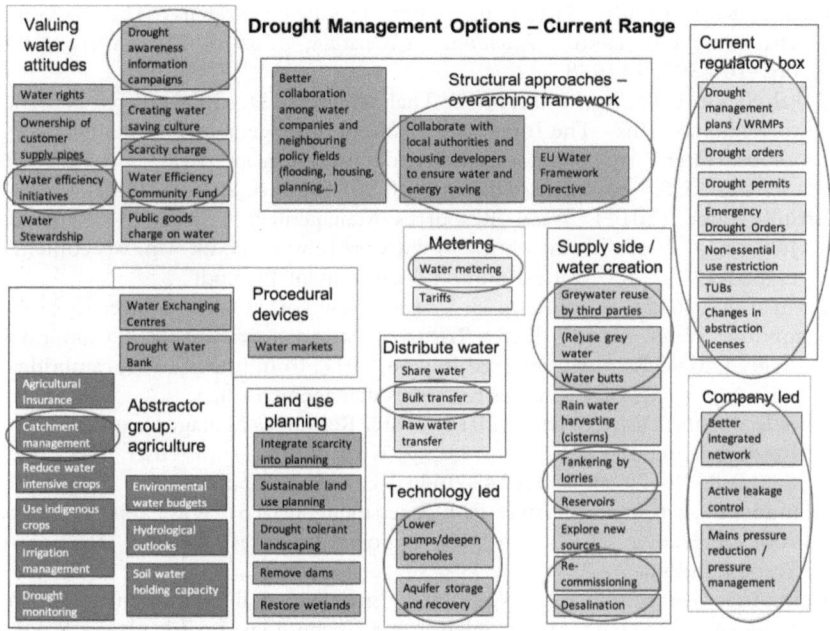

Fig. 2.2 Current range of drought and water scarcity management options in England and Wales (encircled)

REFERENCES

Affinity Water. (2014). Our Plan for Customers & Communities. Final Water Resources Management Plan, 2015–2020. Retrieved July 18, 2017, from https://stakeholder.affinitywater.co.uk/docs/FINAL-WRMP-Jun-2014.pdf.

Albion Water. (2016). Draft Water Resource Management Plan. Retrieved July 18, 2017 from: https://www.albionwater.co.uk/who-we-are/documents-links/useful-documents

Anglian Water. (2014). Water Resources Management Plan 2015. Retrieved July 18, 2017 from: http://www.anglianwater.co.uk/_assets/media/WRMP_2015.pdf

Australian Government. (2017). Agricultural Competitiveness White Paper. Retrieved July 18, 2017, from http://agwhitepaper.agriculture.gov.au/.

Australian Government. National Water Commission (2011) *Water Markets in Australia. A Short History.* Commonwealth of Australia, Canberra.

Bakker, K. J. (2000). Privatizing Water, Producing Scarcity: The Yorkshire Drought of 1995. *Economic Geography, 76*, 4–27. https://doi.org/10.2307/144538.

Bokal, S., Grobicki, A., Kindler, J., & Thalmeinerova, D. (2014). From National to Regional Plans—The Integrated Drought Management Programme of the Global Water Partnership for Central and Eastern Europe. *Weather and Climate Extremes, 3*, 37–46. https://doi.org/10.1016/j.wace.2014.03.006.

Bristol Water. (2014). Water Resources Management Plan 2014. Retrieved July 18, 2017 from: https://www.bristolwater.co.uk/wp/wp-content/uploads/2017/03/Water-Resource-Management-Plan.pdf

Bryman, A. (2012). *Social Research Methods.* New York: Oxford University Press.

Cambridge Water. (2014). *Water Resources Management Plan 2014.* Cambridge Region. Main Report. Retrieved July 18, 2017, from http://www.cambridge-water.co.uk/customers/water-resources-management-plan.

Cholderton and District Water. (2014). Water Resources Management Plan 2014. Retrieved July 18, 2017 from: http://www.sitesplus.co.uk/user_docs/274/File/WRMP-v7%201%202014%20Final_compressed.pdf

Committee on Climate Change Risk Assessment. (2016). *UK Climate Change Risk Assessment 2017.* Synthesis Report: Priorities for the Next Five Years. London.

Committee on Sustainable Water and Environmental Management in the California Bay-Delta, Water Science and Technology Board, Ocean Studies Board, et al. (2012). *Sustainable Water and Environmental Management in the California Bay-Delta.* Washington, DC.

Cook, C. (2016). Drought Planning as a Proxy for Water Security in England. *Current Opinion in Environmental Sustainability, 21*, 65–69. https://doi.org/10.1016/j.cosust.2016.11.005.

Critchley, R., & Marshallsay, D. (2013). *Water Resources in the Southeast.* Progress Towards a Shared Water Resources Strategy in the South East of England Phase 2B Report.

Dee Valley Water. (2013). Water Resources Management Plan December 2013. Retrieved July 18, 2017 from: https://www.deevalleywater.co.uk/wp-content/uploads/2016/07/Executive-Summary-V3-0.pdf

Defra and Environment Agency. (2015). How to Write and Publish a Drought Plan. London, Bristol. Retrieved July 18, 2017 from: https://www.gov.uk/government/collections/how-to-write-and-publish-a-drought-plan

Department of Environment, Food & Rural Affairs, Welsh Assembly Government, Environment Agency. (2011). *Drought Permits and Drought Orders.* Information from the Department of Environment, Food & Rural Affairs, Welsh Assembly Government and the Environment Agency.

Department of Water Resources, Department of Food and Agriculture. (2009). *California's Drought. Water Conditions & Strategies to Reduce Impacts.* Report to the Governor March 30, 2009. Sacramento.

EC. (2000). Directive 2000/60/EC of the European Parliament and of the Council of 23 October 2000 Establishing a Framework for Community Action in the Field of Water Policy.

EEC. (1992). Council Directive 92 / 43 / EEC of 21 May 1992 on the Conservation of Natural Habitats and of Wild Fauna and Flora.

Environment Agency. (2015a). *Drought Response: Our Framework for England*. Bristol.

Environment Agency. (2015b). *Water Company Drought Plan Guideline*. Bristol.

Environment Agency. (2016). *Managing Water Abstraction*. Environment Agency, Bristol

Environment Agency, Natural Resources Wales. (2013). *Water Stressed Areas— Final Classification*. Environment Agency: Bristol; Natural Resources Wales: Cardiff.

Environment Agency, Natural Resources Wales. (2016, May). *Final Water Resources Planning Guideline*. Environment Agency, Bristol.

Essex & Suffolk Water. (2014). Final Water Resources Management Plan 2014. Retrieved July 18, 2017, from https://www.eswater.co.uk/_assets/documents/ESW_Final_Published_PR14_WRMP_Report_-_V3_-_08OCT14.pdf.

Falkenmark, M., Berntell, A., Jägerskog, A., Lundqvist, J., Matz, M., & Tropp, H. (2007). *On the Verge of a New Water Scarcity—A Call for Good Governance and Human Ingenuity*. SIWI Policy Brief. Stockholm: SIWI. Retrieved from https://www.siwi.org/publications/on-the-verge-of-a-new-water-scarcity/.

Farmer, A. M. (Ed.). (2012). Chapter 5.12 Water Scarcity and Droughts. In *Manual of European Environmental Policy*. London: Routledge.

Folke, C., Hahn, T., Olsson, P., & Norberg, J. (2005). Adaptive Governance of Social-Ecological Systems. *Annual Review of Environment and Resources, 30*, 441–473.

Garrote, L., Iglesias, A., Moneo, M., et al. (2007). Application of the Drought Management Guidelines in Spain [Part 2. Examples of Application]. In A. Iglesias, A. López-Francos, & M. Moneo (Eds.), *Drought Management Guidelines Technical Annex* (pp. 373–406). Zaragoza: CIHEAM / EC MEDA Water.

Gleick, P. H., & Heberger, M. (2012). The Coming Mega Drought. *Scientific American, 306*, 14–14.

Gleick, P. H., Christian-Smith, J., & Cooley, H. (2011). Water-Use Efficiency and Productivity: Rethinking the Basin Approach. *Water International, 36*, 784–798. https://doi.org/10.1080/02508060.2011.631873.

Gómez Gómez, C. M., & Pérez Blanco, C. D. (2012). Do Drought Management Plans Reduce Drought Risk? A Risk Assessment Model for a Mediterranean River Basin. *Ecological Economics, 76*, 42–48. https://doi.org/10.1016/j.ecolecon.2012.01.008.

Grecksch, K., & Lange, B. (2019). *Water Efficiency in the Public Sector. The Role of Social Norms. A Primer.* Oxford: Centre for Socio-Legal Studies, University of Oxford.

Green, J. M. H., Cranston, G. R., Sutherland, W. J., et al. (2017). Research Priorities for Managing the Impacts and Dependencies of Business Upon Food, Energy, Water and the Environment. *Sustainability Science, 12,* 319–331. https://doi.org/10.1007/s11625-016-0402-4.

Gupta, J., Pahl-Wostl, C., & Zondervan, R. (2013). 'Glocal' Water Governance: A Multi-Level Challenge in the Anthropocene. *Current Opinion in Environmental Sustainability, 5,* 573–580. https://doi.org/10.1016/j.cosust.2013.09.003.

Hanak, E., Lund, J., Dinar, A., et al. (2011). *Managing California's Water. From Conflict to Reconciliation.* San Francisco: Public Policy Institute of California.

Horne, J. (2016). Water Policy Responses to Drought in the MDB, Australia. *Water Policy, 18,* 28–51. https://doi.org/10.2166/wp.2016.012.

Howitt, R., MacEwan, D., Medellin-Azuara, J., et al. (2015). *Economic Analysis of the 2015 Drought For California Agriculture.* Center for Watershed Sciences, University of California Davis, Davis, CA.

Ingram, B. L., & Malamud-Roam, F. (2013). *The West without Water: What Past Floods, Droughts, and Other Climatic Clues Tell Us About Tomorrow.* Berkeley: University of California Press.

Kampragou, E., Apostolaki, S., Manoli, E., et al. (2011). Towards the Harmonization of Water-Related Policies for Managing Drought Risks across the EU. *Environmental Science & Policy, 14,* 815–824. https://doi.org/10.1016/j.envsci.2011.04.001.

Kron, W., Schlüter-Mayr, S., & Steuer, M. (2016). Drought Aspects—Fostering Resilience through Insurance. *Water Policy, 18,* 9–27. https://doi.org/10.2166/wp.2016.111.

Lange, B., & Cook, C. (2015). Mapping a Developing Governance Space: Managing Drought in the UK. *Current Legal Problems, 68,* 1–38. https://doi.org/10.1093/clp/cuv014.

Lloyd-Hughes, B. (2013). The Impracticality of a Universal Drought Definition. *Theoretical and Applied Climatology, 117,* 607–611. https://doi.org/10.1007/s00704-013-1025-7.

Lorenzo-Lacruz, J., Vicente-Serrano, S. M., López-Moreno, J. I., et al. (2010). The Impact of Droughts and Water Management on Various Hydrological Systems in the Headwaters of the Tagus River (Central Spain). *Journal of Hydrology, 386,* 13–26. https://doi.org/10.1016/j.jhydrol.2010.01.001.

Marsh, T., Cole, G., & Wilby, R. (2007). Major Droughts in England and Wales, 1800–2006. *Weather, 62,* 87–93. https://doi.org/10.1002/wea.67.

Mayring, P. (2008). *Qualitative Inhaltsanalyse. Grundlagen und Techniken* (10th ed.). Beltz: Weinheim.

Met Office. (2012). Dry Weather During 2003. *Met Office*. Retrieved July 28, 2017, from http://www.metoffice.gov.uk/climate/uk/interesting/2003dryspell.html.

Met Office. (2013). England and Wales Drought 2010 to 2012. *Met Office*. Retrieved July 28, 2017, from http://www.metoffice.gov.uk/climate/uk/interesting/2012-drought.

Met Office. (2016). Dry Spell 2004/6. *Met Office*. Retrieved July 28, 2017, from http://www.metoffice.gov.uk/climate/uk/interesting/2004_2005dryspell.

Nelson, R., Howden, M., & Smith, M. S. (2008). Using Adaptive Governance to Rethink the Way Science Supports Australian Drought Policy. *Environmental Science & Policy, 11*, 588–601. https://doi.org/10.1016/j.envsci.2008.06.005.

Northumbrian Water. (2014). Final Water Resources Management Plan 2014. Retrieved July 18, 2017, from https://www.nwl.co.uk/your-home/environment/water-res-man-plan.aspx.

Peel Water Networks. (2013). *Revised Draft Water Resources Management Plan 2013*. Manchester: Peel Water Networks Limited.

Pérez-Urdiales, M., & García-Valiñas, M. Á. (2016). Efficient Water-Using Technologies and Habits: A Disaggregated Analysis in the Water Sector. *Ecological Economics, 128*, 117–129. https://doi.org/10.1016/j.ecolecon.2016.04.011.

Piure, A. (2014). *Markets, Water Shares and Drought: Lessons from Australia. What Can the Water Industry in England and Wales Learn from Australia's Water Reform Story?* Winston Churchill Memorial Trust; Anglian Water.

Portsmouth Water. (2014). Final Water Resources Management Plan 2014. Retrieved July 18, 2017, from https://www.portsmouthwater.co.uk/wp-content/uploads/2015/05/9A3E1C1C-2773-4BBE-B8E2-A0C365AB18F9.pdf.

PPIC. (2015). *Allocating California's Water. Directions for Reform*. Public Policy Institute of California, San Francisco

Priscoli, J. D., & Hiroki, K. (2016). Introduction. *Water Policy, 18*, 1–5. https://doi.org/10.2166/wp.2016.325.

Rijsberman, F. R. (2006). Water Scarcity: Fact or Fiction? *Agricultural Water Management, 80*, 5–22. https://doi.org/10.1016/j.agwat.2005.07.001.

Robins, L., Burt, T. P., Bracken, L. J., et al. (2017). Making Water Policy Work in the United Kingdom: A Case Study of Practical Approaches to Strengthening Complex, Multi-Tiered Systems of Water Governance. *Environmental Science & Policy, 71*, 41–55. https://doi.org/10.1016/j.envsci.2017.01.008.

Rossi, G., & Cancelliere, A. (2013). Managing Drought Risk in Water Supply Systems in Europe: A Review. *International Journal of Water Resources Development, 29*, 272–289. https://doi.org/10.1080/07900627.2012.713848.

Sayers, P. B., Yuanyuan, L., Moncrieff, C., et al. (2017). Strategic Drought Risk Management: Eight 'Golden Rules' to Guide a Sound Approach. *International Journal of River Basin Management, 15*, 239–255. https://doi.org/10.108 0/15715124.2017.1280812.

Sembcorp Bournemouth Water. (2015). Water Resources Management Plan. Final Water Resources Management Plan-2014 Technical Report. Retrieved July 18, 2017 from: http://www.bournemouthwater.co.uk/company-information/economic-regulation/water-resources-plan.aspx

Severn Trent. (2014). Final Water Resources Management Plan 2014. Retrieved July 18, 2017 from: https://www.severntrent.com/content/dam/stw/ST_Corporate/About_us/Docs/WRMP-2014.pdf

South East Water. (2014). Water Resources Management Plan. Retrieved July 18, 2017, from http://www.southeastwater.co.uk/about-us/our-plans/water-resources-management-plan-2014/wrmp-library.

South Staffs Water. (2014). Water Resources Management Plan 2014. Main Report. Retrieved July 18, 2017 from: https://www.south-staffs-water.co.uk/media/1429/ssw_wrmp_2014.pdf

South West Water. (2014). Water Resources Management Plan. Retrieved July 18, 2017 from: https://www.southwestwater.co.uk/globalassets/documents/water_resources_management_plan_june_20141.pdf

Southern Water. (2014). *Water Resources Management Plan 2015–40*. Technical Report. Retrieved July 18, 2017, from https://www.southernwater.co.uk/media/default/pdfs/WRMP-technical-report.pdf.

Speight, V. L. (2015). Innovation in the Water Industry: Barriers and Opportunities for US and UK Utilities. *WIREs Water, 2*, 301–313. https://doi.org/10.1002/wat2.1082.

SSE Water. (2014a). Water Resources Management Plan (England) 2015–2040. SSE Water. Revised Draft Consultation. Retrieved July 8, 2017 from: https://www.sse.co.uk/help/water/resource-management-plan#item1

SSE Water. (2014b). Water Resources Management Plan (Wales) 2015–2040. SSE Water. Draft Consultation. Retrieved July 18, 2017 from: https://www.sse.co.uk/help/water/resource-management-plan#item1

Stakhiv, E. Z., Werick, W., & Brumbaugh, R. W. (2016). Evolution of Drought Management Policies and Practices in the United States. *Water Policy, 18*, 122–152. https://doi.org/10.2166/wp.2016.017.

State of California, Natural Resources Agency, California Department of Water Resources. (2010). *California Drought Contingency Plan*. Sacramento.

Stone, R. C. (2014). Constructing a Framework for National Drought Policy: The Way Forward—The Way Australia Developed and Implemented the National Drought Policy. *Weather and Climate Extremes, 3*, 117–125. https://doi.org/10.1016/j.wace.2014.02.001.

Sutton and East Surrey Water. (2014). Final Water Resources Management Plan. Main Report. Retrieved July 18, 2017 from: http://www.waterplc.com/user-files/file/WRMP_Final_MainReport.pdf

Thames Water. (2014). Final Water Resources Management Plan 2015–2040. Retrieved July 18, 2017 from: https://corporate.thameswater.co.uk/About-us/Our-strategies-and-plans/Water-resources/Our-current-plan-WRMP14

UK Water Act, c. 21. (2014). Retrieved August 4, 2020, from https://www.legislation.gov.uk/ukpga/2014/21/contents.

UK Water Industry Act, c. 56. (1991). Retrieved August 4, 2020, from https://www.legislation.gov.uk/ukpga/1991/56/contents.

United Utilities. (2015). United Utilities Final Water Resources Management Plan. March 2015. Retrieved July 18, 2017 from: https://www.unitedutilities.com/globalassets/z_corporate-site/about-us-pdfs/water-resources/wrmp-mainreport_acc17.pdf

Van Loon, A. F., & Van Lanen, H. A. J. (2013). Making the Distinction between Water Scarcity and Drought Using an Observation-Modeling Framework. *Water Resources Research, 49,* 1483–1502. https://doi.org/10.1002/wrcr.20147.

van Loon, A. F., Gleeson, T., Clark, J., et al. (2016). Drought in the Anthropocene. *Nature Geoscience.* Retrieved February 2, 2018, from https://www.nature.com/articles/ngeo2646.

Van Loon, A. F., Stahl, K., Di Baldassarre, G., et al. (2016). Drought in a Human-Modified World: Reframing Drought Definitions, Understanding, and Analysis Approaches. *Hydrology and Earth System Sciences, 20,* 3631–3650. https://doi.org/10.5194/hess-20-3631-2016.

Veolia Water Projects. (2014). Water Resources Management Plan. Final Published Report. Retrieved July 18, 2017 from: https://www.veolia.co.uk/sites/g/files/dvc636/f/assets/documents/2016/08/5141_WATER_RESOURCES_MANAGEMENT_PLAN.pdf

Vose, J. M., Clark, J. S., Luce, C. H., & Patel-Weynand, T. (2016). *Effects of Drought on Forests and Rangelands in the United States: A Comprehensive Science Synthesis.* Washington, DC: Unites States Department of Agriculture.

Walker, G. (2014). Water Scarcity in England and Wales as a Failure of (Meta) Governance. *Water Alternatives, 7,* 388–413.

Walker, B., Holling, C. S., Carpenter, S. R., & Kinzig, A. P. (2004). Resilience, Adaptability and Transformability in Social–Ecological Systems. *Ecology and Society, 9.*

Water UK. (2016). *Water Resources Long Term Planning 2015–2065.* London.

Welsh Water. (2014). Final Water Resources Management Plan. Technical Report. Retrieved July 18, 2017 from: http://www.dwrcymru.com/en/Environment/Water-Resources/Water-Resource-Management-Plan.aspx

Wessex Water. (2014). *Final Water Resources Management Plan.* Website Version. Retrieved July 18, 2017, from https://www.wessexwater.co.uk/waterplan/.

Wilhite, D. A. (2002). Combating Drought through Preparedness. *Natural Resources Forum, 26,* 275–285. https://doi.org/10.1111/1477-8947.00030.

Wilhite, D. A., Svoboda, M. D., & Hayes, M. J. (2007). Understanding the Complex Impacts of Drought: A Key to Enhancing Drought Mitigation and Preparedness. *Water Resources Management, 21,* 763–774. https://doi.org/10.1007/s11269-006-9076-5.

Wilhite, D. A., Sivakumar, M. V. K., & Pulwarty, R. (2014). Managing Drought Risk in a Changing Climate: The Role of National Drought Policy. *Weather and Climate Extremes, 3,* 4–13. https://doi.org/10.1016/j.wace.2014.01.002.

WMO. (2014). *Atlas of Mortality and Economic Losses from Weather, Climate and Water Extremes (1970–2012).* Geneva: World Meteorological Organisation.

World Meteorological Organization (WMO), Global Water Partnership (GWP) (2014). *National Drought Management Policy Guidelines: A Template for Action (D.A. Wilhite).* WMO, GWP, Geneva, Stockholm.

Yorkshire Water. (2014). Water Resources Management Plan. Retrieved July 18, 2017 from: https://www.yorkshirewater.com/sites/default/files/Water%20Resources%20Management%20Plan%20-%20Introduction%20and%20supply_0.pdf

Zetland, D. (2016). The Struggle for Residential Water Metering in England and Wales. *Water Alternatives, 9,* 120–138.

CHAPTER 3

Drought and Business

Abstract In water resources management and drought and water scarcity management specifically, large business and industrial abstractors are a neglected stakeholder group. Yet, supply chains and production processes often rely on continuous water supply. Droughts and water scarcity can therefore have an adverse effect on businesses and industry. This exploratory study focuses on the UK, where droughts are a recurring feature of the climate, and three business sectors—horticulture, food and drink, and the Scotch whisky industry. The analysis of interviews with trade bodies from the three sectors is complemented by relevant literature and documents. The findings show that the role of drought within each sector differs, how important relationships with other business and industry sectors are, as well as the importance of good relationships with regulatory bodies and water companies. The results highlight that proactive measures such as new water-saving technologies or water recycling are of equal importance.

Keywords UK • Drought • Water scarcity • Business • Industry • Water management

© The Author(s), under exclusive license to Springer Nature Switzerland AG 2021
K. Grecksch, *Drought and Water Scarcity in the UK*, Global Challenges in Water Governance,
https://doi.org/10.1007/978-3-030-65578-5_3

43

Introduction

This chapter seeks to shed light on a new and underexplored issue in water resource management: how large UK water consumers prepare for water shortages, which are becoming increasingly frequent as part of a changing UK climate (Committee on Climate Change Risk Assessment 2016; Marsh et al. 2007) and their relationship with regulatory authorities. The most recent drought event in the UK took place between 2010 and 2012 (Met Office 2013). Drought, although difficult to define precisely (Lloyd-Hughes 2013; Wilhite et al. 2014), is here defined as deficit of water relative to normal conditions (Lloyd-Hughes 2013). Factors that further exacerbate the UK's risk for droughts are climatic changes, population growth and changing water demand patterns. Climate change will intensify the competition for water among agriculture, ecosystems, settlements, industry and energy production, affecting region water, energy and food security (Bates et al. 2008; Jiménez Cisneros et al. 2014). Equally difficult to define is water scarcity. Rijsberman (2006, p. 6) introduces the issue well:

> When an individual does not have access to safe and affordable water to satisfy her or his needs for drinking, washing or their livelihoods we call that person water insecure. When a large number of people in an area are water insecure for a significant period of time, then we can call that area water scarce. It is important to note, however, that there is no commonly accepted definition of water scarcity. Whether an area qualifies as "water scarce" depends on, for instance: (a) how people's needs are defined—and whether the needs of the environment, the water for nature, are taken into account in that definition; (b) what fraction of the resource is made available, or could be made available, to satisfy these needs; (c) the temporal and spatial scales used to define scarcity.

Falkenmark et al. (2007) state that water scarcity not only results from a physical lack of water but is also often a sign of difficulties in mobilising more of the freshwater resources available. Among those difficulties are cost, infrastructure-related challenges and the size of the population. The authors also differentiate between blue water scarcity (water from rivers and aquifers) and green water scarcity (water in the soil for crop production). Similar but simpler, van Loon and van Lanen (2013), for example, define water scarcity as the result of long-term unsustainable use of water resources, which water managers can influence. Walker (2014) stresses that water scarcity is hence human induced and subject to the socio-political and economic context. Bakker (2000) puts the natural, social and discursive elements of water scarcity and puts them into the context of water privatisation in England and Wales. In her example of the Yorkshire drought of 1995, she analyses the drought as the production of scarcity in nature, thereby underlining the influence of socio-economic and other factors, in this case the privatisation of water supply in England and Wales.

While potential impacts of climate change on organisations are discussed in the literature (Gasbarro et al. 2016; Linnenluecke et al. 2012; Weinhofer and Busch 2013; Winn et al. 2011), industrial or large business water abstractors, however, are often less discussed in academic research and policy debates, which are focused on domestic consumption of water (Brown and Hess 2017; Hoolohan and Browne 2016; Nauges and Wheeler 2017) or primarily focused on the agricultural sector (Rey et al. 2016). Lange and Cook (2015) map and discuss the drought governance space for managing drought in the UK and among other important issues name the key actors in the drought governance space. We find the regulatory bodies, water companies, water consultancies, the economic regulator for the water companies (Ofwat), but business and industrial abstractors do not belong to it.

Yet, water is used by businesses and industry either directly for manufacturing products or for washing, cooling or heating during production processes. As early as the 1950s, rising water demands by UK industry became an issue. Water was used in expanding electricity, nuclear power, steel, chemical, and paper industries (Taylor et al. 2009). The 1959 drought highlighted the needs of industries: 'In September 1959, representatives of the heavy and chemical industries of Teesside, reduced to half supply, met with the Tees Valley and Cleveland Water Board, pointing out that this kind of production 'can't just be switched on and off suddenly'. The drought was estimated to have cost Teesside firms around £100,000' (Taylor et al. 2009). During the 1976 drought, the emphasis was on protecting national industry (Morren 1980).

Hence, businesses and industries need water in order to maintain supply chains and production lines, yet water consumers are discussed in a general sense with limited distinction between the public and private sectors, and the different types of large private industrial abstractors who have different needs and organisational histories in dealing with water. According to the latest report by the UK's Committee on Climate Change, 'some water intensive industries are clustered in areas at risk of water scarcity such as paper manufacturing in Kent and chemicals manufacturing in the northwest of England' (Committee on Climate Change Risk Assessment 2016). The impact of flooding is usually more visible not least supported by the media (BBC News 2016), yet the effects and consequences such as interrupted supply chains potentially are the same for droughts and water scarcity. Upholding supply chains and production processes is not only necessary from a wider economic point of view, but especially interruptions to food and drink supply chains can have knock-on effects such as empty supermarket shelves or increased prices for groceries. Industries and businesses should therefore have an intrinsic interest in proactive drought and water scarcity management in order to avoid the potential consequences.

In Spain, the losses of the 2007–2008 drought were significant. Apart from agriculture, sectors such as horticulture, hydropower production and non-market welfare losses like environmental damage (river basin ecosystems), all suffered from the consequences of water scarcity. 'The total costs of the extreme drought event affecting the metropolitan area of Barcelona in 2007 and 2008 have been estimated here at 1,605 million Euros' (Martin-Ortega et al. 2012). Jenkins (2013) emphasises the indirect losses of drought, for example mining, retail, finance and insurance, educational services, healthcare, arts, entertainment, recreation or accommodation and food services. Linnenluecke et al. (2012) point out with reference to the 2010 Russian drought and wildfires that although companies in the affected areas have significant adaptive capabilities for competitive environments, they were unable to assemble or did not possess suitable resources and capabilities for withstanding and recovering from impacts.

For the UK, Benton et al. (2012, p. 15) highlight the sensitivity of the UK food supply chain: 'The UK food chain is inherently resilient due to the diversity of suppliers, retailers and sources of imported food. However, the just-in-time nature of the UK food chain does mean that severe weather may cause significant disruption, albeit this likely to be over a timescale of days, or at most a week or two in the case of severe weather.' More specifically for a certain sector, Thomas (2015, p. 194) points out, 'Clearly, climate change is going to affect barley quality as well as yield, which will in turn affect the efficiency of major end-users such as maltsters, brewers and distillers.'

This chapter aims to answer the question what role drought and water scarcity have among UK industries and businesses and whether these industries and businesses have strategies and plans to first of all react to drought and water scarcity and second if they already apply any proactive measures to prevent potential disruptions from drought and water scarcity. Furthermore, this research tries to elicit the relationship businesses and industries have to the regulatory bodies and water companies. This chapter not only focuses on individual companies but rather tries to explore a general picture based on policies and strategies by trade associations. In addition, this chapter presents three sectors, food and drink, horticulture and the Scotch whisky industry, in detail. This chapter therefore has limitations; however, it contributes to the rich debate about water governance and water resources management and also to discussions about the so-called water-energy-food nexus (WEF-nexus) with focus on businesses (e.g. Green et al. 2017; 2030 Water Resources Group 2020).

MATERIALS AND METHODS

The data presented in this chapter rests on two main sources: a literature and document review, and a small number of exploratory semi-structured expert interviews. The aim of the literature and document review was to provide an overview over current ideas, strategies and policies regarding large water abstractors and drought and water scarcity. The review included first of all academic literature. The review of academic literature focused on the UK and the European Union. Academic literature was searched using Scopus and Web of Science search engines. Articles were selected on the basis that they deal with industries or businesses and drought and water scarcity and a snowball search using cross-references. The document review included documents from UK trade associations that could be affected heavily by drought and water scarcity. This included the paper industry, food and drink industry, Scotch whisky industry, the horticultural trade, the British glass industry, and the British hydropower industry. The websites of these trade organisations were searched for documents, policies or statements with regard to drought and water scarcity. The results of both, the literature and document review, are presented and discussed in the next section.

In the next step, representatives of all mentioned trade associations were contacted via email or telephone and interviews requested. Only three trade associations responded positively, and interviews were conducted with the Horticultural Trade Association (HTA), the Food and Drink Federation (FDF) and the Scotch Whisky Association (SWA) (Table 3.1). Others did either not respond or decline the interview request.

Table 3.1   List of interviews

Interview name	Interviewee
Interview HTA	John Adlam (Horticultural Trade Association)
Interview FDF	David Bellamy (Food and Drink Federation)
Interview SWA	Both interviewees anonymised (representative from trade association and representative from whisky distillery)

Those who declined mentioned that the issue is not seen as relevant among their membership. The semi-structured interviews lasted one hour and were conducted based on methods of the focused, problem-centred and expert interview (Fowler and Mangione 1990; Merton and Kendall 1946; Witzel 2000). The collected literature and documents and transcribed interviews were analysed based on the method of qualitative content analysis (Bryman 2012; Mayring 2008). Given the limited number of interviews, this study is exploratory and it does not claim any representativeness. However, it provides a useful snapshot into key industry sectors who are heavily relying on constant freshwater supply. The focus in this chapter is especially on the relationship of businesses and industry with regulatory agencies, hence the focus on trade association instead of individual businesses. For example, the UK Food and Drink Federation (FDF) represents over 7000 businesses in the UK. Hence, getting the view on drought and water scarcity issues from the representative trade association can be taken as an indicator for the views and actions of the sector.

In order to answer the research question, the following categories were used to analyse the data. First, the role of drought and water scarcity within trade organisations was assessed. This tried to elicit the importance of the issue and whether it is competing against other water-related issues such as flooding or wastewater recycling or any non-water-related issues. Second, the analysis assessed in how far trade associations' membership businesses could be affected by drought and water scarcity and whether this is reflected in any strategies, policies or reports. The interruption of supply chains and the availability of water for production processes are just two of the potential consequences of drought and water scarcity, and this category tries to establish in how far trade associations are aware of possible affects. A further category assessed the nature and extent of the relationship between trade associations and regulatory bodies such as the UK Department for the Environment, Food and Rural Affairs (Defra), the Environment Agency (EA), the Scottish Environment Protection Agency (SEPA) and UK water companies. These actors represent the key actors in the UK drought governance space (Lange and Cook 2015). Hence, relationships with them are vital for effective drought and water scarcity management. A fourth category elicited the experience with previous drought events and whether this resulted in any changes of the approach towards drought and water scarcity. Since drought is a recurring feature of the UK climate (Marsh et al. 2007) and recent drought events took place between 2010 and 2012, 2004 and 2006, 2003, and 1995 and 1996 (Met Office 2012, 2013, 2016), UK businesses and industries were affected in the past, and this category tries to determine how they responded to these events. A last category assessed what future challenges but also solutions trade associations identify and envisage with regard to drought and water scarcity. This future-oriented question elicited any strategies, instruments or technologies to be better prepared for future drought and water scarcity events. To some extent this overlaps with previous categories, but the question was asked specifically at the end of the interview to allow respondents to think broadly with regard to drought and water scarcity.

In summary, these categories try to explore key aspects of how UK business and industry trade associations approach drought and water scarcity as an issue. The analysis does not claim to provide a comprehensive picture but a first explorative insight supported by examples into a neglected aspect of not only drought and water scarcity management but water resources management in general.

Results and Discussion

This section starts with a general overview based on the review of academic literature and documents. It will then present results based on the analysis of trade association documents and the three expert interviews. The section concludes with a summarising discussion of the results.

Kurland and Zell (2010) declare a 'paucity of water-related studies' in top journals in this regard. In relation to planetary boundaries and business, Whiteman et al. (2012) conclude that businesses have not addressed water scarcity despite its importance. Ya He and Cranston (2014) demonstrate that businesses choose collaborative solutions if they recognise that poor water security is a threat to their businesses. Their example demonstrated that once interdependencies were identified by the different stakeholders and collaboration recognised, more effective water strategies and financial mechanisms were delivered that recognised the value of water to the interdependent nexus elements across different sectors (cited in Green et al. 2017). Green et al. (2017) discuss the critical role of the private sector around food, energy, water and the environment under the so-called Water-Energy-Food (WEF) nexus. This 'nexus' approach highlights the interdependencies between water, energy and food. From the perspective of water, water is not seen as a sector or issue area but as a cross-cutting issue, which requires changes in governance in all relevant sectors. 'This is on the basis of the argument that water governance cannot itself regulate land, agriculture or other issues in society but that water issues need to be taken into account in each of these governance processes' (Gupta et al. 2013, p. 577). Green et al.'s (2017) study identifies 40 key questions that, if answered, help companies understand and manage their water-food-energy-environment nexus dependencies and impacts. A common issue among a number of questions was problem awareness or in the words of the authors 'understanding how risks manifest around unsustainable management of food, energy, water and environmental systems is key for businesses operating under conditions of increased demand for natural resources' (Green et al. 2017). The 2030 Water Resources Group, an initiative by the World Bank Group, also highlights the need to act now to minimise the future impact of water scarcity on human wellbeing, social cohesion and economic development (2030 Water Resources Group, p. 11). In the following, the results of the three expert interviews are described and discussed based on the categories as outlined in the Materials and Methods section.

Role of Drought and Water Scarcity

The role of drought and water scarcity differs from sector to sector. It depends, for example, on whether water is used during the production process, afterwards, for cleaning purposes, for instance, or both. The Horticultural Trade Association (HTA), who represents the interests of the gardening industry, sees the role of drought and water scarcity as critical. 'Container grown plants have no connection to the soil and require frequent irrigation. Although modern growing media have a fairly grained structure by design, high water holding capacity etc., many of them grow in glass houses with no natural rain. In a drought most businesses would be out of business within 24 hours' (Interview HTA).

The Food and Drink Federation (FDF) represents the interests of the UK food and drink industry, the largest manufacturing sector in the country. It is a diverse sector with over 400,000 employees and around 7000 businesses across the UK (Food and Drink Federation 2017). The FDF believes it is important to distinguish between a genuine drought emergency and ongoing water scarcity issues in some parts of the country brought about by pressures from climate change and increasing water demand. 'Normally in drought situations the Environment Agency would ask for voluntary 'hands off' reductions in abstraction first with mandatory reductions very much a last resort' (Interview FDF). Separately, there has been much discussion recently within the sector regarding the government's proposals to move forward with abstraction licence reform in England and how to ensure that water for food and drink production is accorded the right priority within any new system.

The Scotch Whisky Association (SWA) represents over 90% of the Scotch whisky industry. The industry considers itself to be in a fortunate position with regard to drought and water scarcity. 'Scotland is lucky in terms of available water' (Interview SWA). About 80% of the water that is used for Scotch whisky production is cooling water, that is water temperature will likely be the more limiting factor rather than water quantity. Generally speaking, the Scotch whisky industry is very thoughtful about water, because it is one of the key ingredients or critical resources in a whisky (Interview SWA). In addition, most of the distilleries are situated in rural places, away from industrial-intensive areas; therefore, conflicting water pressures are less of an issue.

Potential Effects of Drought and Water Scarcity and Reflection of Drought and Water Scarcity in Strategies, Policies or Reports

The potential effects can vary between short interruptions to production processes and complete disruptions of supply chains, which can stall production all together. It was mentioned above that in case of a drought many horticultural businesses would be out of business within 24 hours due to their reliance on irrigation water. 'Landscape gardening is severely constrained in a drought event, hose pipe bans affect garden centres and this could have serious financial disadvantages for a business with a multi-billion pound turnover' (Interview HTA). However, larger nurseries may have the room and the space to provide a supply buffer, for example an on-site reservoir or a recycling scheme.

The impact for FDF businesses would be great, because 'nearly all food processing businesses need a constant supply of water, 24/7' (Interview FDF). Hence, any variation or interruption of supply would be damaging for production. Therefore, within the FDF membership, there has been much emphasis on the importance of water efficiency in operations and building resilience within supply chains (ibid.). In a report, 'Every Last Drop – Saving water along the food supply chain' (Food and Drink Federation 2011), the FDF tries to encourage companies to use water more efficiently and gives tips how to harvest water, save costs and be aware of water use. It also contains a suggestion to implement an effective environmental management system (Food and Drink Federation 2011).

The effect on Scotch whisky production would be a slowing down of the production process according to the SWA: 'Effects are in terms of temperature, or cooling water availability, which is slowing production' (Interview SWA). During a drought event in Scotland in 2008, there were some incidents where Scotch whisky production had to be stopped in some places for short periods (Interview SWA). The SWA uses Environmental Strategy and Progress reports (Scotch Whisky Association 2015) to highlight where water is coming from and what it is used for. Their environmental strategy also highlights issues such as climate change and sets water efficiency targets. One of the two interviewees, a representative from a whisky distillery, said that they engage actively with water issues, for example, through a water risk study, which assesses water in a given geographical location and includes social, regulatory and physical aspects of actual water availability, the activity of stakeholders and regulatory pressures (Interview SWA). In combination with work on climate models, assessing water resilience, this results in a water risk strategy, an assessment of regulatory pressures, climate pressures and plan on how to deal with these issues.

Relationship with Regulatory Bodies and Water Companies

As mentioned in the introduction, industry and businesses are not part of the key stakeholders in the UK drought governance space. Nonetheless, the relationship with regulatory bodies and water companies is vital before, during and after a drought. The HTA is member of the Abstraction Reform Advisory Group, a body that was set up by the environment ministry Defra to carry out a reform of the abstraction licencing scheme. In addition, they are members of the Water for Food Group, a collaboration overseen by the Environment Agency that includes everyone who is related to water, food and agricultural processing. This includes, for example, plant and tree nurseries, the Agriculture and Horticulture Development Board (AHDB), the potato industry, chartered surveyors, irrigators and drainage boards. The group discusses and produces a range of documents at government level. In the East Anglia region, the HTA is also involved in the Water Resources Anglia group alongside stakeholders such as Anglia Water, the National Farmers Union and Cranfield University.

The FDF has good links with both Defra and the Environment Agency on water-related issues (Interview FDF). It chairs the EA's sector liaison group which meets twice a year. In addition, FDF is a member of the coalition of agri food interests known as the Water for Food Group, which also includes the HTA and the National Farmers Union.

The key relationships the SWA holds are with Defra, Scottish Water and the Scottish Environment Protection Agency (SEPA). The relationship with water companies is on an ad hoc basis because many distillers have their own private water supply. 'We have a very good and proactive relationship with SEPA' (Interview SWA), which even sees the Scotch whisky industry as a best practice example on how SEPA would like to work with other business sectors in Scotland. SEPA's One Planet Prosperity regulatory approach uses a sectoral approach. 'Sector plans will focus on practical ways of delivering environmental, social and economic outcomes. They will specify existing levels of compliance, the market context for that sector and the key issues faced by the sector and SEPA' (SEPA 2016). The distillery representative mentioned the membership of the Beverage Industry Environment Round table (BIER), where water is one of the topics they discuss.

Experience with Past Drought Events and Changes in the Approach Towards Drought and Water Scarcity

There have been recent drought events in the UK, for example 2010–2012 (Met Office 2013), 2004–2006 (Met Office 2016) or 2003 (Met Office 2012). The HTA attended the drought summit that was convened by Defra after the 2010–2012 drought. The interviewee himself remembered the 1976 drought. 'The last drought event immediately galvanised the mind and we went to several water companies to ask them to introduce a traffic light scheme [indicating the severeness of drought using green, yellow and red]. The water companies' reaction was that they agreed to look at it' (Interview HTA). Another issue that was picked up after the last drought event was to look at more drought-hardy plants such as cacti or succulents; however, they are not as colourful as other ornamental plants. 'The horticultural market is a leisure market, not a food market' (Interview HTA).

The FDF also mentions the 2010–2012 drought and the good communication it was able to establish with its members as a result of the information provided by the Environment Agency.

As mentioned before the effect of drought or water scarcity on Scotch whisky production is minimal as opposed to other industries. The SWA recalls a few days without or reduced production for a handful of locations over the last decade. The 2012 drought caused furthermore short periods of low flows (Interview SWA). As a response we find the above-mentioned water risk study by the distillery as well as the updated Environmental Strategy and Progress reports by the SWA and its collaboration with SEPA.

Future Challenges and Solutions

For the HTA, one of the key strategies is to think about water-saving processes: 'There are some very sophisticated technologies, for example looking at soil moisture content, or telling the volumetric water content of a media and if it goes down a certain level, irrigation is triggered' (Interview HTA). Other solutions include the use of rain gauges that stop irrigation if it rains, the increased use of smart metering and the use of more recycled water: 'We could use more grey water' (ibid.). In a wider perspective, the HTA mentions the use of canals to move water across areas and to hold rainfall back instead of letting it run off into the sea. Another option would be aggregate abstraction licences in an area with many plant nurseries and take the water from where the most is available, linking nurseries by pipes.

For the FDF one of the major future challenges will be how to find ways to increase national food production to meet national priorities around economic growth including exports and food security given the pressures on water availability in some parts of the country (Interview FDF). In a separate report by Defra and the EA (2016), a beverage company is showcased. Here, a review on water usage has prompted changes in run time on pumps and other equipment. The manufacturing site has become more self-sufficient also due to its own on-site water source.

The strategy for the SWA is to use bigger companies to take leadership and share knowledge, technology and experience. 'The big companies' strategy gives us a framework to have these discussions and if there are any lessons to learn in order to be ahead of the game' (Interview SWA). From a distiller's perspective, they are fortunate to be in a resilient water catchment and that they are the only ones there. 'We do rely on the regulatory framework to keep it that way. (…) If land management changes, planned forestry or less planned development that could change water availability' (Interview SWA).

Other Industries and Large Abstractors

The above-detailed analysis focused on three sectors, and the objective was to provide an exploratory insight into drought and water scarcity and its management among large business and industrial abstractors in the UK. As mentioned in the introduction and the methods section, there are, of course, many more water-intensive UK industries such as the cement industry, chemical industries, paper industries, mineral industries, hydropower, the energy sector, especially thermal power plants or the glass industry. They were contacted for an interview, but most of them declined or they did not respond at all to requests. Defra has also produced a study on the impacts of drought in England on various sectors in 2013 (DEFRA 2013). The following therefore presents an overview of documents that are publicly available.

The UK's paper-based industries are unified under the umbrella of the Confederation of Paper Industries (CPI). They represent 70 companies, among them makers of soft tissue papers, paper and board manufacturers, corrugated packaging producers and many more. Their aim is to promote paper as a sustainable and renewable material and to reach the best regulatory and legislative practice. The industry has £6.5bn annual turnover capital and more than 100,000 employees (CPI 2017a). In its annual review the CPI (2017b) writes, 'Turning closer to home, issues around water scarcity and indeed water abundance abound.' The annual review also highlights the importance of water scarcity and the idea/need of adaptation to the challenges of climate change in general. The paper and pulp sector has its own guidance to prepare for climate change supported by the Environment Agency (2013). This action plan contains a questionnaire to assess businesses, recommendations for creating a risk map and how to explore critical thresholds of water usage.

According to the British Hydropower Association, hydropower in the UK generates almost 6000 GWh electricity per year (British Hydropower Association 2017). Hydropower generation relies to 100% on water availability, or in other words, drought and water scarcity could bring hydropower generation to a complete halt.

The Construction Products Association states in a report that 'Whilst concern is usually on water scarcity, the increasingly volatile weather patterns in the UK are causing regular, often severe, flooding (five of the wettest years on record have been since 2000); for manufacturers both scarcity and flooding can pose a business risk' (Thornback et al. 2015). The report also highlights the different uses of water in various construction product manufacturing processes such as pre-cast concrete, plastics, wood panels, ceramics, glass, quarries and steel (Thornback et al. 2015). Citing a member survey, they say that manufacturers are becoming increasingly aware of water-related issues. 'Business risks can manifest in two main ways – either through restriction of the quantity of water available (whether physical or regulatory related) or though declining water quality' (Thornback et al. 2015, p. 12).

Summary

There is no one-size-fits-all solution to how businesses and industry in the UK should deal with drought and water scarcity. Responses, strategies and measures depend on business and industry size, geographical location, the interconnectedness of the supply chain, government policies and what water is used for predominantly—for the product or service itself, as coolant or for cleaning. All interviewed and analysed sectors have had experiences with droughts and water scarcity in the recent past, and all of them are aware that future drought and water scarcity events can cause disruption to businesses and industries. Noteworthy is the position of the SWA, who, although water scarcity and droughts are not an imminent threat to them, takes care of water issues proactively because it is one of the key ingredients to their product. What unites all three interviewed sectors is the importance of relationships with other stakeholders, predominantly neighbouring sectors, regulatory bodies and water companies. This is, for example, obvious in the case of the Food and Drink sector. Many ingredients are sourced from the agricultural sector, and it is therefore vital to collaborate with, for example, the National Farmers Union. The Water for

Food group overseen by the Environment Agency is another positive example that fosters collaboration among the different stakeholders.

The HTA even proactively engaged with water companies after the last drought in order to be better prepared for future droughts. In the East Anglia region, the HTA approached water companies, Essex and Suffolk Water and Affinity Water, to introduce a traffic light scheme so that people and businesses can better prepare for a drought event (Interview HTA). Another approach is to focus on technological solutions, either, with reference to the sectors covered in this study, at distillery or plant nursery level. Water recycling and the use of recycled water, drought-hardy plants, on-site reservoirs, canals to distribute water among nurseries in vicinity to each other or soil moisture measurement technologies that only trigger irrigation when really necessary are measures to adapt to varying water availability but are also measures to increase water efficiency.

This exploratory study gave a snapshot of how different business and industry sectors in the UK approach drought and water scarcity, and the scope of the study is therefore limited. As mentioned in the Materials and Methods section, many more industries were contacted but either declined an interview or did not respond at all. Those who declined mentioned that the issue is not relevant for their membership or not discussed by their membership. This could be addressed by raising awareness among industries, for example, by linking the issue to other relevant water-related issues such as flooding. Water scarcity often is a consequence of flooding. Since flooding in general is usually more appealing to stakeholders, this could be an entry point to widen the debate with drought and water scarcity issues. Nonetheless, the selected sectors use water in their operations and production processes in different ways and do exemplify some of the major challenges that are incurred by drought and water scarcity.

It is important to emphasise that drought and water scarcity are only two of many other possible extreme weather events that could affect businesses and industry. Flooding, heat waves, landslides, bush and forest fires or hurricanes can disrupt supply chains and production processes (Linnenluecke et al. 2012).

Conclusions

The aim of this chapter was to explore how different business and industry sectors in the UK deal with drought and water scarcity. Not only in the light of climatic changes do drought and water scarcity have ecological consequences and put risk on domestic users of water, but they also hold severe implications for businesses, industries and their supply chains ranging from slowed-down production to the interruption of production processes over long time periods. The role of drought and water scarcity and the extent to how each sector could be affected unsurprisingly varies across the different sectors. Nonetheless, businesses and industries in the UK are aware of the potential effects of drought and water scarcity either because they experienced drought events before or they are forward thinking in their approach to water resources. The solutions to future challenges of drought and water scarcity are tackled through technological solutions and cooperation with other sectors and regulatory bodies. Especially the latter should be fostered and institutionalised as businesses and industries currently do not belong to the key stakeholders in the UK drought governance space. For example, the Water Resources in the South East Group (WRSE), the Water Resources East Group in Anglia and most recently the Water Resources North (WReN) are three organisations in the UK that foster the collaboration between water companies, regulators and other stakeholders in the respective regions. They could serve as a model for other regions as well.

References

Bakker, K. J. (2000). Privatizing Water, Producing Scarcity: The Yorkshire Drought of 1995. *Economic Geography, 76*, 4–27. https://doi.org/10.2307/144538.

Bates, B. C., Kundzewicz, Z. W., Wu, S., Palutikof, J. P. (2008). *Climate Change and Water*. Technical Paper of the Intergovernmental Panel on Climate Change. IPCC, Geneva.

BBC News. (2016). Ginger Nut Biscuit 'Crisis Over.' *BBC News*.

Benton, T., Gallani, B., Jones, C., et al. (2012). *Severe Weather and UK Food Chain Resilience*. Detailed Appendix to Synthesis Report. Food Research Partnership: Resilience of the UK Food System Subgroup, Swindon.

British Hydropower Association. (2017). Hydro in the UK. Retrieved June 13, 2017, from http://www.british-hydro.org/hydro_in_the_uk.

Brown, K. P., & Hess, D. J. (2017). The Politics of Water Conservation: Identifying and Overcoming Barriers to Successful Policies. *Water Policy, 19*, 304–321. https://doi.org/10.2166/wp.2016.089.

Bryman, A. (2012). *Social Research Methods*. New York: Oxford University Press.

Committee on Climate Change Risk Assessment. (2016). *UK Climate Change Risk Assessment 2017*. Synthesis Report: Priorities for the Next Five Years. London.

CPI. (2017a). Who We Are. Retrieved May 26, 2017, from http://www.paper.org.uk/aboutcpi/pages/who_we_are.html.

CPI. (2017b). *Pride in Paper. Review 2016/17*. Confederation of Paper Industries, Swindon.

DEFRA. (2013). *The Impacts of Drought in England*. R&D Technical Report WT0987/TR. Department for Environment, Food and Rural Affairs, London.

Department of Environment, Food and Rural Affairs, Environment Agency. (2016). *Creating a Better Place. Our Ambition to 2020*. Department for Environment, Food and Rural Affairs; Environment Agency, London, Bristol.

Environment Agency. (2013). *Preparing a Climate Change Action Plan: Paper & Pulp Sector Guidance*. Environment Agency.

Falkenmark, M., Berntell, A., Jägerskog, A., Lundqvist, J., Matz, M., & Tropp, H. (2007). *On the Verge of a New Water Scarcity—A Call for Good Governance and Human Ingenuity*. SIWI Policy Brief. Stockholm: SIWI. Retrieved from https://www.siwi.org/publications/on-the-verge-of-a-new-water-scarcity/.

Food and Drink Federation. (2011). *Every Last Drop—Saving Water Along the Food Supply Chain*. Food and Drink Federation.

Food and Drink Federation. (2017). FDF—Stats at a Glance. Retrieved August 5, 2017, from https://www.fdf.org.uk/statsataglance.aspx.

Fowler, F. J., & Mangione, T. W. (1990). *Standardized Survey Interviewing. Minimizing Interviewer-Related Error*. Newbury Park: Sage.

Gasbarro, F., Rizzi, F., & Frey, M. (2016). Adaptation Measures of Energy and Utility Companies to Cope with Water Scarcity Induced by Climate Change. *Business Strategy & the Environment* (John Wiley & Sons, Inc), 25, 54–72. https://doi.org/10.1002/bse.1857.

Green, J. M. H., Cranston, G. R., Sutherland, W. J., et al. (2017). Research Priorities for Managing the Impacts and Dependencies of Business Upon Food, Energy, Water and the Environment. *Sustainability Science, 12*, 319–331. https://doi.org/10.1007/s11625-016-0402-4.

Gupta, J., Pahl-Wostl, C., & Zondervan, R. (2013). 'Glocal' Water Governance: A Multi-Level Challenge in the Anthropocene. *Current Opinion in Environmental Sustainability, 5*, 573–580. https://doi.org/10.1016/j.cosust.2013.09.003.

Hoolohan, C., & Browne, A. L. (2016). Reframing Water Efficiency: Determining Collective Approaches to Change Water Use in the Home. *British Journal of Environment & Climate Change, 6*, 179–191.

Jenkins, K. (2013). Indirect Economic Losses of Drought under Future Projections of Climate Change: A Case Study for Spain. *Natural Hazards, 69*, 1967–1986. https://doi.org/10.1007/s11069-013-0788-6.

Jiménez Cisneros, B. E., Oki, T., Arnell, N. W., et al. (2014). Freshwater Resources. In C. B. Field, V. R. Barros, D. J. Dokken, et al. (Eds.), *Climate Change 2014: Impacts, Adaptation, and Vulnerability. Part A: Global and Sectoral Aspects.* Contribution of Working Group II to the Fifth Assessment Report of the Intergovernmental Panel of Climate Change. Cambridge University Press, Cambridge, United Kingdom and New York, NY, USA, pp. 229–269.

Kurland, N. B., & Zell, D. (2010). Water and Business: A Taxonomy and Review of the Research. *Organization & Environment, 23,* 316–353. https://doi.org/10.1177/1086026610382627.

Lange, B., & Cook, C. (2015). Mapping a Developing Governance Space: Managing Drought in the UK. *Current Legal Problems, 68,* 1–38. https://doi.org/10.1093/clp/cuv014.

Linnenluecke, M. K., Griffiths, A., & Winn, M. (2012). Extreme Weather Events and the Critical Importance of Anticipatory Adaptation and Organizational Resilience in Responding to Impacts. *Business Strategy & the Environment* (John Wiley & Sons, Inc), 21, 17–32. https://doi.org/10.1002/bse.708.

Lloyd-Hughes, B. (2013). The Impracticality of a Universal Drought Definition. *Theoretical and Applied Climatology, 117,* 607–611. https://doi.org/10.1007/s00704-013-1025-7.

Marsh, T., Cole, G., & Wilby, R. (2007). Major Droughts in England and Wales, 1800–2006. *Weather, 62,* 87–93. https://doi.org/10.1002/wea.67.

Martin-Ortega, J., González-Eguino, M., & Markandya, A. (2012). The Costs of Drought: The 2007/2008 Case of Barcelona. *Water Policy, 14,* 539–560. https://doi.org/10.2166/wp.2011.121.

Mayring, P. (2008). *Qualitative Inhaltsanalyse. Grundlagen und Techniken* (10th ed.). Beltz: Weinheim.

Merton, R. K., & Kendall, P. L. (1946). The Focussed Interview. *American Journal of Sociology, 51,* 541–557.

Met Office. (2012). Dry Weather During 2003. *Met Office.* Retrieved July 28, 2017, from http://www.metoffice.gov.uk/climate/uk/interesting/2003dryspell.html.

Met Office. (2013). England and Wales Drought 2010 to 2012. *Met Office.* Retrieved July 28, 2017, from http://www.metoffice.gov.uk/climate/uk/interesting/2012-drought.

Met Office. (2016). Dry Spell 2004/6. *Met Office.* Retrieved July 28, 2017, from http://www.metoffice.gov.uk/climate/uk/interesting/2004_2005dryspell.

Morren, G. E. B. (1980). The Rural Ecology of the British Drought of 1975–1976. *Human Ecology, 8,* 33–63. https://doi.org/10.1007/BF01531467.

Nauges, C., & Wheeler, S. A. (2017). The Complex Relationship Between Households' Climate Change Concerns and Their Water and Energy Mitigation Behaviour. *Ecological Economics, 141,* 87–94. https://doi.org/10.1016/j.ecolecon.2017.05.026.

Rey, D., Holman, I. P., Daccache, A., et al. (2016). Modelling and Mapping the Economic Value of Supplemental Irrigation in a Humid Climate. *Agricultural Water Management, 173*, 13–22. https://doi.org/10.1016/j.agwat.2016.04.017.

Rijsberman, F. R. (2006). Water Scarcity: Fact or Fiction? *Agricultural Water Management, 80*, 5–22. https://doi.org/10.1016/j.agwat.2005.07.001.

Scotch Whisky Association. (2015). *Scotch Whisky Industry*. Environmental Strategy Report 2015. Edinburgh.

SEPA. (2016). *One Planet Prosperity—Our Regulatory Strategy*. (Scottish Environment Protection Agency).

Taylor, V., Chappells, H., Medd, W., & Trentmann, F. (2009). Drought Is Normal: the Socio-Technical Evolution of Drought and Water Demand in England and Wales, 1893–2006. *Journal of Historical Geography, 35*, 568–591. https://doi.org/10.1016/j.jhg.2008.09.004.

Thomas, W. T. B. (2015). Drought-Resistant Cereals: Impact on Water Sustainability and Nutritional Quality. *Proceedings of the Nutrition Society, 74*, 191–197. https://doi.org/10.1017/S0029665115000026.

Thornback, J., Snowdon, C., Anderson, J., & Foster, C. (2015). *Water Efficiency. The Contribution of Construction Products*. London: Construction Products Association.

Van Loon, A. F., & Van Lanen, H. A. J. (2013). Making the Distinction between Water Scarcity and Drought Using an Observation-Modeling Framework. *Water Resources Research, 49*, 1483–1502. https://doi.org/10.1002/wrcr.20147.

Walker, G. (2014). Water Scarcity in England and Wales as a Failure of (Meta) Governance. *Water Alternatives, 7*, 388–413.

Weinhofer, G., & Busch, T. (2013). Corporate Strategies for Managing Climate Risks. *Business Strategy & the Environment* (John Wiley & Sons, Inc), 22, 21–144. https://doi.org/10.1002/bse.1744.

Whiteman, G., Walker, B., & Perego, P. (2012). Planetary Boundaries: Ecological Foundations for Corporate Sustainability. *Journal of Management Studies, 50*, 307–336. https://doi.org/10.1111/j.1467-6486.2012.01073.x.

Wilhite, D. A., Sivakumar, M. V. K., & Pulwarty, R. (2014). Managing Drought Risk in a Changing Climate: The Role of National Drought Policy. *Weather and Climate Extremes, 3*, 4–13. https://doi.org/10.1016/j.wace.2014.01.002.

Winn, M., Kirchgeorg, M., Griffiths, A., et al. (2011). Impacts from Climate Change on Organizations: A Conceptual Foundation. *Business Strategy & the Environment* (John Wiley & Sons, Inc) 20, 157–173. https://doi.org/10.1002/bse.679.

Witzel, A. (2000). The Problem-Centered Interview. *Forum Qualitative Social Research, 1*, Art. 22.

Ya He, J., & Cranston, G. R. (2014). *The Cambridge Natural Capital Leadership Platform. Sink or Swim: A Multi-sector Collaboration on Water Asset Investment.* Cambridge, UK: Institute for Sustainability Leadership.

2030 Water Resources Group. (2020). *Building Trust, Growing Resilience.* Washington, DC: World Bank Group.

PART II

Knowledge

CHAPTER 4

Drought and Drought Knowledge: The Example of Catfield Fen

Abstract This chapter introduces specific results from a larger study on how environmental science knowledge is related to regulatory tools for managing drought and water scarcity. The focus is on the quality of hydroecological data, the role of local (expert) knowledge and power relationships between actors. These three themes will be introduced in detail and their implications showcased in the example of Catfield Fen. Catfield Fen, a wetland nature reserve, is a prime example of how the influence of a particular specific hydroecological data set exerts regulatory influence and also shows how access to a very specific data set can shape power relationships between actors.

Keywords Drought • Water scarcity • Knowledge • Abstraction • Catfield Fen

INTRODUCTION

It cannot be said often enough, drought is a recurring feature of the United Kingdom (UK) climate (Marsh et al. 2007). While Chap. 2 introduced the general framework of drought and water scarcity in the UK and how to manage it, the previous chapter, with its focus on large businesses and industrial abstractors, made clear that drought and water scarcity are

© The Author(s), under exclusive license to Springer Nature
Switzerland AG 2021
K. Grecksch, *Drought and Water Scarcity in the UK*, Global
Challenges in Water Governance,
https://doi.org/10.1007/978-3-030-65578-5_4

not just about water companies and providing water supply. Water is used in production processes, for cleaning or for cooling. The extent to which water shortages will occur in the future in the context of a changing climate depends on the way in which water resources are governed and regulated, also on the basis of scientific knowledge about water availability (Gupta et al. 2013; Jacobs et al. 2016; Pahl-Wostl et al. 2020) and 'in order to be effective, knowledge systems that support decisions about water-resource management and development must link research- and experience-based knowledge to practices across a broad range of challenges' (Jacobs et al. 2016). The consideration of the various sources of knowledge has emerged as an essential component of thinking about environmental and resource governance (Charles et al. 2020). The link between knowledge and decision-making is not fixed, but rather semipermeable, moveable and negotiable (Jasanoff and Wynne 1998). New knowledge is acquired; old knowledge is discarded. Hence this arrangement needs flexibility and translation between different types of knowledges, for example scientific knowledge and lay knowledge (Cash et al. 2003). Ingram (2013) emphasises that knowledge content refers to what is being understood, and this includes scientific knowledge, experiential knowledge and traditional knowledge. Anderson et al. (2019) urge the wider acceptance and inclusion of diverse knowledge of, for example, rivers. Tengö et al. (2014) highlight the need to promote dialogue among different knowledge systems for improved policy and to support decision-making. The challenge is to balance the breadth and depth of the various forms of knowledge from various groups; however, including all viewpoints, voices and interests may lead to an inefficient process of decision-making and knowledge production. 'Yet, excluding viewpoints, voices and interests from dialogues will result in power imbalances' (Earth System Governance Project 2018, p. 38). Lave (2015) concludes that we are not seeing the decline of scientific authority (over knowledge) but its redistribution, as a far wider range of knowledge producers are accorded intellectual legitimacy. She cautions that while the democratisation of knowledge production can be regarded as progress, one also needs to take into account that if an environmental non-governmental organisation can produce knowledge so can corporations thereby pursuing their commercial knowledge interests. The World Social Science Report 2016 (ISSC et al. 2016, p. 22) speaks of knowledge inequalities, which 'includes the question of whose knowledge counts and what types of knowledge are considered most important. Knowledge inequalities between individuals and groups affect the capacity to make informed decisions, to access services and to participate in political life.' The example presented in this chapter highlights such a knowledge inequality. Moreover, Hulme et al. (2020) plead for a diversification of knowledge and to abolish knowledge hierarchies, where certain forms of knowledge dominate or are seen as superior. Therefore, 'because the process by which knowledge is produced and communicated determines, to a large extent, its quality and usability, evaluating an actor's useable knowledge should involve examining not just what they know or think they know, but how they came to know it' (Williams et al. 2015, p. 83).

This chapter takes a closer look at how drought and water scarcity related knowledge is generated in England and Wales, by whom and how it is used. Hence, the main research question explores how environmental science knowledge informs the use of regulatory tools for managing drought and water scarcity in England and Wales. With regard to environmental science knowledge and regulatory tools, this chapter discusses three key themes resulting from interviews with water companies, regulators, consultancies and abstractor groups. First, the quality and validity of the data, especially hydroecological data, generated; second, the power relationships between the different actors who are involved in drought and water scarcity management; and third, the role of local (expert) knowledge and the scope for greater consideration of local expert knowledge of catchments and their response to drought. Finally, this chapter illustrates these three themes and how highly contested environmental science knowledge can inform the mobilisation of regulatory tools with reference to the example of a review of agricultural abstraction licences at the Catfield Fen wetland nature reserve. The Catfield Fen example provides an example of how the influence of a particular specific hydroecological data set, in this case about the rare fen orchid, exerts regulatory influence and also shows how access to a very specific data set can shape relationships of power. In the Catfield Fen example, the data weakened the farmer's case as an abstractor whose abstraction rights were curtailed on the basis of this very precise information.

The context of this chapter is a larger study that aimed to identify the main environmental science knowledges that inform the use of regulatory tools in relation to drought and water scarcity management, in particular in England and Wales, and, in addition, sought to explore how a specific range of environmental science knowledges shape relationships of power between key institutional actors in the drought governance space. The study was part of MaRIUS Drought project (see Chap. 2), and a full report (Grecksch and Lange 2018) is available online from the about-drought.info website. The original study has a much larger scope and refers to three individual case studies with regard to environmental science knowledge and regulatory tools: drought planning, the Restoring Sustainable Abstraction (RSA) process and historic droughts in the UK. The study generated primary data and developed a number of regulatory tools and environmental science knowledges and also identified themes and knowledge gaps based on a large set of interviews with water companies, regulators, consultancies and abstractor groups. This chapter takes some of the key themes identified and relates them to the example of Catfield Fen to illustrate the key issue of how environmental science knowledges inform regulatory tools and what this means for the power relationship between the different actors in the UK drought governance space.

For the further discussion it is important to define the key concepts used in this chapter. The first key concept is 'environmental science knowledge(s)'. This describes specific types of expertise that identify causes, incidences and impacts of drought and water scarcity. It includes a range of knowledges that characterise the state of the natural environment from a science perspective and that are deployed in practice by water companies, regulators or consultants when a proposal is made for a particular regulatory tool to be deployed, for example a drought order. Examples of environmental science knowledges are data on the availability of water in reservoirs or the building and application of hydrological models of surface and groundwater flows.

However, there is variation in the degree to which these knowledges are clearly defined with reference to particular methodologies, and there is variation in the degree to which they are required to be used when regulatory decisions are taken. Some environmental science knowledges such as Strategic Environmental Assessments (SEA), Habitats Regulation Assessments (HRA) or Water Framework Directive (WFD) assessments are required by law and further defined by legal provisions. Other environmental science knowledges are developed by water companies and regulatory agencies themselves, sometimes in an ad hoc way, in order to support internal decision-making, being less directly required or defined by law.

The second important concept is 'regulatory tool'. Regulatory tools are measures that are deployed by either a water company or a state regulator, such as the Environment Agency for England (EA) or Natural Resources

Wales (NRW), in order to prevent the occurrence of water scarcity or alleviate drought conditions during a drought. Examples are Temporary Use Bans (TUBs), drought permits and drought orders. Water companies can implement Temporary Use Bans that restrict watering your garden or washing your private motor vehicle, under their own powers. These restrictions are temporary measures that are intended to reduce the demand for water and are usually one of the first steps a water company will take to protect its supplies during a drought. Another regulatory tool that water companies can deploy is drought permits. These are authorised by the EA and enable a water company to increase the supply of water abstracted from the natural environment. A drought order on the other hand, issued by the Secretary of State heading Defra (Department for Environment, Food & Rural Affairs) or the Welsh Minister, can increase abstraction from the environment by water companies in order to meet statutory duties for public water supply. It can also restrict demand from commercial water users or limit abstraction by a water company, other abstractors or the EA.

The next section describes the data and methods. Then three key themes with regard to environmental science knowledges and regulatory tools are discussed. This is followed by a detailed discussion of the Catfield Fen example, highlighting these key themes in a practical context.

Data and Methods

As mentioned in the introduction, this chapter builds on a larger study on the governance of drought and water scarcity in the UK, especially with regard to environmental science knowledges and regulatory tools (Grecksch and Lange 2018). The data basis is a set of 50 qualitative semi-structured interviews with regulators (Defra, EA and Natural England), representatives from English and Welsh water companies, consultancies as well as individual abstractors and abstractor groups, carried out between 2015 and 2017. Also, as mentioned above, the data were collected for three case studies—drought planning, the Restoring Sustainable Abstraction (RSA) process and the use of regulatory tools during specific historic drought episodes, in particular the 2010–2012 (Met Office 2013) and 2003–2004 (Met Office 2012) droughts in the UK. The drought planning case study examined the use of environmental science knowledges during the drafting phase of drought plans, while the RSA case study analysed how environmental science knowledge(s) inform decisions by regulatory agencies to modify or revoke abstraction licences. The idea behind the RSA programme was the revision of abstraction licences with the aim to promote a fairer distribution of water in catchments, thus enhancing abstractors' support for specific water allocations, which can become highly contested during an actual drought. The third case study examined the use of environmental science knowledges for mobilising specific regulatory tools during particular historic drought episodes. We will see that the RSA process provides the majority of data for the Catfield Fen example, which was further enriched by the analysis of regulatory decisions, legal provisions, stakeholder reports and grey literature. The interviews asked for the environmental science data used and the relationship among the key stakeholders such as water companies and the Environment Agency.

The interview data was analysed using NVivo 11 to identify themes. For the Catfield Fen example, a case study approach was employed. A case study entails the detailed and intensive analysis of a single case and is concerned with the complexity and particular nature of the case in question (Bryman 2012). According to Yin (2009), 'it is an empirical enquiry that investigates a contemporary phenomenon in depth and within its real-life context when the boundaries between phenomenon and context are not clearly evident.' Finally, a note on interview quotation: interviews are anonymised and quoted in a standardised format. For example, 'RSA1.WC2' stands for Restoring Sustainable Abstraction case study interview, Water Company 2. 'CON' stands for consultant, 'REG' for regulator, 'ABS' for abstractor (group) and 'OTH' for other. 'DP' stands for the Drought Planning case study.

Three Key Themes

The following section presents and discusses three key themes resulting from the analysis of the interviews. Many more themes were identified (Grecksch and Lange 2018); however, the three themes presented here were selected due to their relevance for the Catfield Fen example. Their significance goes beyond the example though. Themes are recurring ideas, issues or statements expressed by interviewees often not necessarily as a direct answer to questions from the interview schedule. Themes can as well be identified from the study of literature and documents. Therefore, identifying themes can help to discover further dimensions of drought and water scarcity management.

Importance of Hydroecological Data and the Need for More and Better Hydroecological Data

Hydroecological data cover the interaction between water and wildlife habitats of a catchment delivering an integrated view of drought risk and thus enabling a more holistic approach to drought management. Thus, besides standard monitoring data like river and ground water flows, integrating hydroecological knowledge into drought planning provides stakeholders with more and better data and knowledge to make informed decisions. The importance of generating and integrating more hydroecological data into drought management was expressed across the range of interviews, that is water companies, consultants, regulators but also abstractor groups. At the same time many of them also acknowledged the current lack of such data.

> From my point of view, understanding how resilient different rivers are to different levels of drought and human pressure on top of the droughts. That's much more ecological research I think, but I think it's very useful to understand that a bit better so water companies and regulators can have more confidence in that. (Interview DP1.CON1)

One water company stressed that environmental resilience and what is happening with river ecology in the future are key areas that need more research (Interview DP1.WC12). Another water company also emphasised the importance of knowledge about environmental state baselines: 'When we go into a drought, it's all very well throwing loads of money at it then and looking at it but if you've got nothing to compare it to, it's pretty worthless. So, I'd like to see a bit more the environment aspect pushed a bit further than just as an afterthought' (Interview DP1.WC4). The view from abstractor groups is that they would like to see more research into knowledge about soil, water and irrigation (Interview RSA.ABS1). The Environment Agency states that it would like to see more research around the environmental monitoring side, suggesting that their guidance is currently not particularly clear with regard to what they want companies to provide in advance of a drought permit so that they can understand what damage it could do (Interview DP1.REG5).

Mentioned specifically by many interviewees was the Environmental Flow Indicator (EFI), a specific type of environmental science data, which is used to indicate where abstraction pressure may start to cause an undesirable effect on river habitats and species. The EFI is a percentage deviation from the natural river flow represented using a flow duration curve. The flow duration curve is a plot that shows the percentage of time that flow in a stream is likely to equal or exceed some specified value of interest. Flow duration curves can be made for a given river over two different time periods to illustrate if/how the range of flows has changed over time. This percentage deviation is different at different flows. It is also dependant on the ecological sensitivity of the river to changes in flow. The EFI is calculated within the Resource Assessment and Management (RAM) framework. This assessment gives an indication where and when water is available for new abstractions. Flow standards for the Water Framework Directive (WFD) developed by the UK Technical Advisory Group have been adapted to set the EFI (Environment Agency 2013).

Hence, the EFI is set with reference to expert opinion and at a level to support 'good ecological status', the legal standard required to be achieved under the EU Water Framework Directive. The EFI is used in Catchment Abstraction Management Strategies (CAMS) where resource availability is expressed as surplus or deficit of water resources in relation to the EFI. The EFI is also used in the hydrological classification for the WFD to identify water bodies where reduced river flows may be causing or contributing to a failure of 'good ecological status'. This is called a compliance assessment (Environment Agency, 2013). For the EA, the EFI constitutes an 'ecological protection line' (Interview RSA.REG2).

Some stakeholders are critical of the EFI though. One interviewee from a water company mentions that the information that goes into a model is important and getting that information right is crucial:

> [Q]uite often they're [river flows] limited to maybe just one gauging station at the bottom end of a catchment and it hasn't necessarily got all of the information you need for all the upstream parts. (…) I think they're probably the critical pieces and that's the bit we sometimes struggle with and have discussions around, about how much confidence do we have in these numbers that are being generated, because quite often we're producing something, this environmental flow indicator, that's a bit of a made up number and actually how sensitive is that river to those flows? (Interview RSA.WC2)

Another member of staff from a water company suggested that they could often show the EFI was wrong:

It's set at the wrong value in some places because it's based on no or very little data. Generally speaking, the EA wants more data to understand water companies' options: (…) I think they welcome the fact that we do all this information gathering of data. The more information we get, the better. They use our information, as I said before, and put it in their models. So we've collected lots of flow data and we've shown the EFI might be wrong and challenged the EFI, they will use our information. (Interview RSA.WC8)

A different water company expressed a similar opinion:

You can get these statements by the EA that 50% of rivers are over-abstracted and that may be true against the EFI standards but not true once you've investigated them and shown that actually for half of those rivers, they're not as sensitive to flow as you thought when you applied the EFI which is just purely applying the EFI standards. (Interview RSA.WC4)

The EFI is another way of indicating where abstraction can cause an undesirable effect on river habitats and species. Yet, in a wider context this is an example of another fundamental issue emerging from the data: discussing the usefulness of the EFI raises the question whether regulating by 'numbers'—as a form of bureaucratic regulation—solves a resource problem for the regulator and thus enhances its regulatory control. If there are, however, only limited data underpinning instruments such as the EFI, then it may actually limit regulatory control and legitimacy.

The significance of hydroecological data should thus not be underestimated and was mentioned and stressed by a large number of interviewees. Reference to the significance of hydroecological data also helped to identify a gap in drought and water scarcity management, that is a lack of these data.

The desire to generate and integrate more hydroecological data into drought governance was expressed across the range of interviews. Water companies stressed environmental resilience and what is happening with river ecology in the future as key areas that need more research: 'We know on the Wye, the salmon run, there will be a point at which the salmon will not run and it won't because there's not enough water in the river, it'll be because of the temperature. So, there's probably quite a lot to do there in understanding how the ecology might change and what the ecology needs in the future' (Interview DP1.WC12).

If more and better hydroecological data is to be included in decision-making in relation to drought and water scarcity in the future, the question arises who—regulators, water companies or both—should collect the data. Monitoring, the most important environmental science knowledge in the UK and a 'quasi regulatory tool' (Grecksch and Lange 2018), is fragmented among the different actors in the drought governance space. Thus, also with regard to relationships of power in the drought governance space, who collects hydroecological data becomes an important question and a governance challenge.

Power Relationship Between Actors

How the responsibilities for generating environmental science knowledges are allocated shapes relationships of power between the key organisational actors in the governance space for drought and water scarcity (Fig. 4.1). For example, shared monitoring responsibilities shift the balance of power from regulatory bodies to water companies and consultancies. The latter provide environmental science knowledge in case water companies lack the necessary in-house capacity to generate it. In order, however, for drought plans or drought permits/orders to be approved, water companies have to provide the necessary data to support their applications, and the way they make a case in these for those regulatory tools. This, though

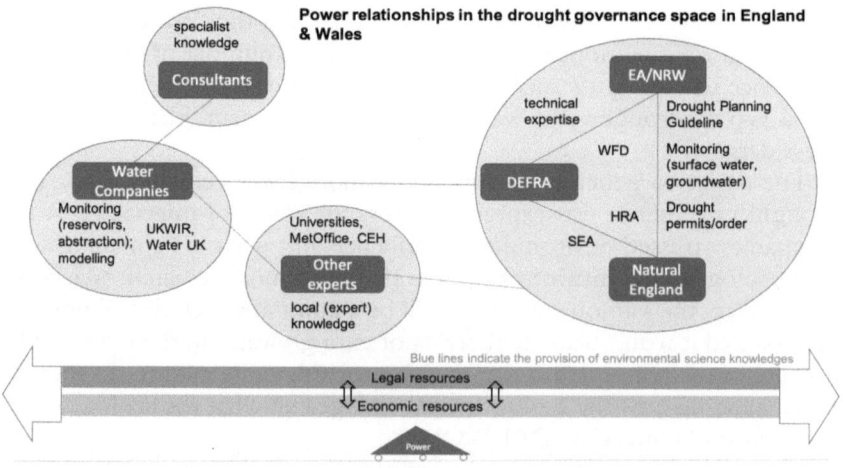

Fig. 4.1   Power relationships in UK drought and water scarcity management

by some water companies described as 'ticking boxes', retains power for regulatory agencies to scrutinise drought management by water companies. Consultants can play an influential role in these relationships of power shaped also by access to data, also in light of the fact that also the Environment Agency sometimes employs consultancies to generate environmental science data. It would therefore be interesting to see how in the case of, for example, more hydroecological data being gathered, this would tilt power relationships among the key stakeholders. Other factors that influence the power relationships between the actors are legal and economic resources. Legal resources refer to primary and secondary legislation that grant legal powers and impose legal duties upon regulatory agencies and regulated organisations and thereby shape relationships of power between actors in the drought governance space. Economic resources refer to financial means of regulatory agencies and regulated organisations in drought governance, for example, to implement particular drought management options, which are also shaped by decisions of the economic regulator of water companies, Ofwat (Water Services Regulation Authority).

Local (Expert) Knowledge

Local (expert) knowledge could be potentially significant in decision-making. Besides the abstract and formal natural science knowledge generated by key organisational stakeholders—water companies, regulatory agencies and consultancies—local, including expert, knowledge can also be important. Local knowledge is knowledge generated and provided—on the basis of personal interests and observations—by local people, for example inhabitants of a catchment. Local expert knowledge is knowledge generated and provided by semi-professional and professional bodies, such as local environmental non-governmental organisations or angling clubs. Local expert knowledge also includes knowledge generated by experts who in their capacity as professionals working either for a regulatory body or water company have accumulated extensive local knowledge, for example, about a catchment or a certain stretch of a river.

Local knowledge can fill gaps left by more abstract and formal environmental science knowledges and add new perspectives. Moreover, including local knowledge in the application of regulatory tools for preventing and managing drought and water scarcity can empower stakeholders and strengthen the legitimacy of regulatory decisions. The analysis suggests, however, that so far, local knowledge remains largely 'untapped'. In the course of the RSA programme, however, many water companies appraised its benefits: 'One of the things about these investigations is it's become clear while we're doing it, is to involve other stakeholders because there's a lot of information out there that's useful information to us that we want to use as well' (Interview RSA.WC8). The same water company now involves, for example, Rivers Trusts and Wildlife Trusts who provide them with useful information as part of RSA investigations: 'We didn't know any of that just looking at maps and the [Environment] Agency didn't know a lot of it either. So, it has really been useful involving these stakeholders and we'll involve them through the whole process' (Interview RSA.WC8).

Natural England blames a lack of resources for missing local knowledge of sites, because at those sites that are National Nature Reserves they have officers out on site every day gathering knowledge which they lack for a lot of other sites (Interview RSA.REG4). 'I think in a drought, you do rely on local expertise,' concludes the EA, emphasising the fact that you rely on people who know how a river, reservoir or groundwater reacts to water scarcity (Interview DP1.REG2). A consultant mentioned that similar limits to local expertise within the EA are also due to the EA's practice of moving its staff around. This leads to recruitment problems and an increasing lack of local knowledge (Interview DP1.CON3). This view is supported by local abstractor groups who also emphasise the value of local knowledge. They see local knowledge as a basis to sustainable abstraction, with local knowledge generated by people who watch borehole levels go up and down over the years or who have seen a chalk stream dry out and regenerate a few times (Interview RSA.ABS1).

The theme 'local (expert) knowledge' shows, first of all, that there is a continuum of knowledge about drought and water scarcity, ranging from anecdotal, often unsystematic local knowledge to knowledge generated by local professionals, for example officers in National Nature Reserves. As Lange and Cook (2015) have shown, the drought governance space in England and Wales is rather confined with regard to the range of stakeholders involved and, as this report shows, at times also with reference to the range of environmental science knowledges and regulatory tools relied upon. Local (expert) knowledge can therefore be a valuable addition to the body of knowledge about drought and water scarcity mobilised in the prevention and management of drought and water scarcity in the UK. Open questions with regard to local knowledge, however, remain such as how to deal with strongly biased views and how to systematise local knowledge, which so far is produced outside of the formal drought governance space.

EXAMPLE: CATFIELD FEN

Catfield Fen is a wetland nature reserve in the county of Norfolk. It is located approximately 15 miles northwest of Great Yarmouth and 23 miles northeast of Norwich and part of the Broads National Park. It is one of the most important areas of fen in the UK providing habitat to many rare species, especially invertebrates but also rare plants such as the fen orchid. Catfield Fen is not open to the public. It is partly owned by Butterfly Conservation, a large butterfly conservation organisation, owning 24 ha (Butterfly Conservation 2018). The remainder is owned privately by the Catfield Hall Estate.

The Catfield Fen example highlights what environmental science knowledges were used in this case, how they link to the application of regulatory tools and to showcase challenges but also opportunities with regard to what knowledge is applied and which stakeholder provides which type of knowledge. The chosen example involves the contestation of environmental science knowledge, thereby also shedding light on power relationships between key actors in the drought governance space. The example will be introduced, different stakeholder views will be discussed and the section will conclude with a summarising discussion. The appendix of this chapter contains the unabridged stakeholder views on the issue.

Catfield Fen and the Implications for Environmental Science Knowledges

In the context of Restoring Sustainable Abstraction process, Catfield Fen is a good example of contentious environmental science knowledge and how it is generated and contested. So far it represents the longest controversy over an abstraction site in England. Two abstraction licences were up for renewal, yet site managers, local ecologists, the Broads Authority (BA) and Natural England raised doubts over the hydrological modelling carried out by the Environment Agency. Opponents claimed it is insufficient to conclude that it will have no effect on the integrity of the site (BAWAG 2018). The affected farmer exercised his right for public inquiry, which he subsequently lost. He also lost his appeal against the decision in September

2016 (Case 2016). The public inquiry concluded that the fen is in danger due to ecological change, increasing acidification is the cause and that water abstraction is the most likely explanation by reducing the flow of alkaline groundwater to the fen (Environment Agency 2015).

The Catfield Fen example concerns one of the primary sources of environmental science knowledges in drought and water scarcity management: modelling (in this case EA's Northern East Anglia Chalk [NEAC] groundwater model). Several stakeholders challenged the applied groundwater model and found it to be inadequate in this particular instance. Originally, the Environment Agency granted an extension of the licences. However, the licences came under reconsideration after new information emerged provided by the landowner. This was concerning the model used by the EA and water quality issues. The latter addresses the assumption that the water abstraction by the farmer is pulling away alkaline groundwater and that the fen is benefitting from less alkaline water but more acidic rainfall. That impact is changing the pH of the soil on the fen, which is having an adverse impact on the population of Fen Orchids.

In 2008 the Environment Agency undertook groundwater modelling to assess if the fen is drying out, as claimed by the landowner. The EA concluded that the site is not drying out, but there is ecological change in the form of acidification caused by a reduction in base-rich groundwater. The landowner and other organisations such as the Broads Authority subsequently provided the EA with more and new information leading to the decision to revoke the two abstraction licences based on the grounds that the EA cannot ascertain beyond reasonable scientific doubt that the abstractions will not adversely affect the site integrity (Environment Agency 2015). Core criticism arose regarding the applied groundwater model and its inadequacies. The Broads Authority expressed concerns regarding the model in response to the EA's groundwater report: 'An inadequate groundwater model development process has been followed; failing to use the Environment Agency modelling guidelines; (…); Given the shortcomings of the conceptual models and computational modelling, the results from the modelling are not reliable and should not be used for licence determination' (Kelly 2014, p. 8).

Natural England concluded that there are instances where new data becomes available or the groundwater model did not pick up the sensitivities of the site (Interview RSA.REG4). The regional water company, who are affected by this through an in-combination effect (see below), wonder where it is taking them:

because it is fundamentally questioning what's happening in terms of all this groundwater modelling and where we've got to previously, and that's my concern and I think the EA's concern on that as well. (…) It does not take long before you can start unravelling a model and say that's where I'm concerned. It's the best tool that we have available now, do we really want to be unpicking these things? I am not comfortable with where we could end up, but I know that is where the challenges are coming in at the moment. (Interview RSA.WC2)

The Catfield Fen example also touches upon how legal principles, such as the precautionary principle, may shape the generation and use of environmental science knowledges. The precautionary principle states that the burden of proof for the proposition that an action is not harmful falls on those taking that action (Knill and Liefferink 2007; Fisher et al. 2019). In other words, in the case of Catfield Fen, the farmer who wanted his two licences renewed had to prove that his abstraction for agricultural purposes does not cause harm to the Catfield Fen site with regard to water quality issues. According to abstractor groups, individual farmers usually lack the resources or the knowledge to oppose regulatory action on biological grounds and that this requires usually significant sums of money (Interview RSA.ABS3).

> A possible cause of the revocation of those licences in Norfolk was the fact that the guy [the landowner of Catfield Estate] who wanted them revoked was able to finance as many different reports and researchers and solicitors and lawyers as he could care to do, because he's always got an unlimited pit of money where that particular subject was concerned. So, he could do that but the farmers, who are on notice really, were unable to provide evidence of their own to refute what's been put forward because they just hadn't got the resources to do it, or to fight the case. (Interview RSA.ABS1)

Furthermore, the example incurred an in-combination effect because not only the farmer applied for the renewal of his two existing licences but Anglia Water also hold a large abstraction licence (of right) in the same area. According to the local water company:

> We've had some letters and we've reached an agreement with the EA to say that as an interim measure, you haven't flagged this soon enough for us to put a solution in place now, we will do something in AMP7 and as for AMP6 we will reach an interim solution that we try and reduce our abstraction on

Ludham [Catfield Fen site], recognising that you've had these challenges from elsewhere. (Interview RSA.WC2)

Summary of the Example

In summary, the Catfield Fen example shows how the use of a key environmental science knowledge, in this case modelling, informs a decision about an abstraction licence. It also shows that the decision relied significantly upon this particular environmental science knowledge. The originally applied model was heavily disputed by several stakeholders and declared as inadequate. However, further local expert knowledge, an overall key theme identified by the research, was provided and influenced the final decision. In addition, Catfield Fen is a good example of the value of hydroecological data. The data that was used to contest the existing data was essentially hydroecological data that provided a more holistic view of the Catfield Fen site and the effect that a continued water abstraction could have. With regard to relationships of power in the drought governance space, the Catfield Fen example is unusual as it only involves water companies marginally. Instead, relationships of power between a well-resourced landowner who was able to commission expertise as opposed to a farmer with fewer resources were a key feature of the particular Catfield Fen example. The example, however, does show how multiple expertise by mostly non-regulatory bodies can challenge an environmental science knowledge basis gathered by one of the main regulatory bodies in the drought governance space. The validity of the EA's groundwater model was questioned, a key environmental science knowledge and usually a strong lever for decisions made by the EA.

CONCLUSION

The aim of this chapter was to demonstrate the importance of environmental science knowledge, especially hydroecological data, local (expert) knowledge and power relationships between actors for the management of drought and water scarcity in the UK. These three themes were discussed and highlighted in the example of Catfield Fen, a case that showcases the link between the application of regulatory tools and the challenges but also opportunities with regard to what knowledge is applied and which stakeholder provides which type of knowledge. It also shed light on power relationships between key actors in the drought governance space.

This power relationship revealed an imbalance regarding access to knowledge. Referring back to the World Social Science Report (ISSC et al. 2016) mentioned above, knowledge inequalities include the question of whose knowledge counts and what types of knowledge are considered most important. In the case study it was the knowledge produced on behalf of the landowner who could afford to pay for external expertise while the affected farmer could not. Independent of whether the actual decision may have been right, one wonders whether a coordinated process of knowledge production, thereby diversifying the knowledge base and including different voices (Tengö et al. 2014), would have led to an agreeable result for all parties involved. Instead this is an example of how knowledge can be paid for by those who possess financial resources, that is money plus knowledge equals power. The landowner was able to finance a process of private knowledge production to pursue his own interests.

On a grander scale, and this may also be interesting for the context beyond the case study, the initial objection was made against the groundwater model applied by the Environment Agency, the environmental regulator for England and authority over the regulations. However, as the representative from the water company said, where do we end up if we can start unravelling the groundwater models. Of course, a groundwater model can be wrong and other parties should have the right to object, but it touches upon a larger point of who has authority and power over knowledge. As Hukkinen (2020, p. 1) writes: 'In contested science-policy interactions, the actors' awareness—or lack of it—of when knowledge performs its power is a key determinant of policy outcomes'. For instance, hydroecological data is crucial. Monitoring data feeds models and the modelling results inform decisions regarding water demand and supply. The general desire expressed by stakeholders is to generate and integrate more hydroecological data. But the results also show that data and knowledge can be contested as in the case of the EFI. However, if more data is to be gathered the questions arise by whom? This has implications for the power relationships between actors as well. Hence, will the Environment Agency collect more data or will it delegate this to the water companies, who can also delegate it to specialised consultancies? The balance in Fig. 4.1 could easily tilt towards private actors, and the regulatory authority, in this case the EA, could find itself in a situation where they no longer hold the majority of hydroecological data, but still set the framework for the modelling. Water companies, consultancies and other private actors could come up with their own model based on the data they have and challenge the authority of the regulatory agency. The concentrated ownership of English water companies could facilitate this step further. A third important element discussed in this chapter is the role of local (expert) knowledge. Local knowledge can fill gaps left by more abstract and formal environmental science knowledges and add new perspectives. And, including local knowledge in the application of regulatory tools for preventing and managing drought and water scarcity can empower stakeholders and strengthen the legitimacy of regulatory decisions. However, so far local knowledge remains a mostly untapped source of knowledge in the UK drought and water scarcity management context.

APPENDIX

The following stakeholder views on the issue are extracted from the interviews as well as other publicly available material. While the preceding paragraphs summarised and discussed the main issues, the purpose of reproducing the different stakeholder views in details is to provide the full and unfiltered account of the discussion for the interested reader.

Regulators

Interview RSA.REG5
That's a licence renewal so we're aware of it happening at Catfield Fen. It's not a case that the Secretary of State would normally deal with but we know what's going on, yes but it's a renewal case so the licence has come to an end and whether they should be renewed. I think there's a public inquiry on that in April. We, as DEFRA would not go along to, the Inspector acts as the Secretary of State and the inspector is the chair of the public inquiry. The Secretary of State can recover appeals, so there is the opportunity for the Secretary of State to ask the Inspector to report to the Secretary of State but that's at the Secretary of State's discretion.

Environment Agency (2015)
The abstraction licence at Ludham Road (Catfield Fen site) was originally granted in 1988 and the Plumsgate Road licence in 1986. From circa 2000 both licences were renewed on a short-term basis, for periods of between two and five years, pending the outcome of the EA's 'Review of Consents' (RoC). In September 2008, the landowner at Catfield Hall (Units 11 and 35 of the Ant Broads and Marshes Site of Special Scientific

Interest [SSSI], component SSSI of the Broads Special Area of Conservation [SAC], Broadland Special Protection Area [SPA] and Broadland Ramsar) expressed concern that Catfield Fen was drying out. In December 2011 the applicant applied to renew both the Plumsgate Road and Ludham Road abstraction licences on the same terms.

In order to understand the concerns about the drying out of Catfield Fen (endorsed by Natural England [NE] in 2011), the Environment Agency (EA) agreed to commission a report through Amec. In recognition of the contentious issue surrounding local abstraction, the EA has treated the applications as being of high public interest.

Following the Amec report, the EA undertook a groundwater modelling assessment using the EA's Northern East Anglia Chalk (NEAC) groundwater model. Through further consultation with NE, the EA were advised that the site was not drying out, but there is ecological change—more *Sphagnum* sp. moss on the site, which indicates that the site is acidifying. The cause of this could be a reduction in base-rich groundwater input into the Fen.

The EA prepared its Appropriate Assessment in March 2014. This was supported by a technical report, with both reports focusing on the impacts of abstraction of the Ant Broads and Marshes component SSSI of the Broads SAC.

The EA received extensive comments from both NE and Broads Authority (BA) through the Appropriate Assessment consultation, including new information. Given the new information, and uncertainties raised by NE and the BA, the EA revised the conclusions of the Appropriate Assessment.

A public consultation was conducted, between 17 November 2014 and 15 December 2014. At the same time, NE and the BA were re-consulted on the statutory assessments.

In order to take account of relevant information received through the public consultation, an extended consultation response date was agreed with NE of 16 January 2015. This enabled NE to consider the updated statutory consultation together with the relevant public consultation responses. NE again provided extensive comments on the updated statutory assessment.

Conclusion

The EA cannot ascertain beyond reasonable scientific doubt that the abstractions will not adversely affect the site integrity of the Ant Broads and Marshes SSSI component of the Broads SAC; Broadland SPA and Broadland Ramsar, given the current fully licensed in-combination level of abstraction. This is specifically in relation to the predicted in-combination water level change at Snipe Marsh and predicted alone and in-combination water chemistry change at Catfield Fen. It is also considered that the Plumsgate Road and Ludham Road abstractions could potentially damage the Ant Broads and Marshes SSSI when considering potential changes in water chemistry. It is therefore recommended that the applications are refused.

Interview RSA.REG6

Yes, the farmer had a time limited licence and it came up for renewal and when it came up for renewal, lots of heads were scratched and a lot of work was done by the Environment Agency and Natural England and a local landowner as well, looking at the impacts of that. It is an interesting case in that it is a relatively small abstraction and it comes back to this in combination issue. When you have a relatively small abstractor but that is on top of public water supply abstraction and some of the smaller abstraction in that area; so it very much hinged around this in combination issue and how you apply that in the Habitats Regulations. So it was that side of things. Exactly how that is going to influence what goes forward with Abstraction Reform, I do not know and we won't really know until we get the outcomes of the public enquiry.

(…)

I wasn't involved directly in the public enquiry so I can't tell you exactly how things went there, but I do know that the Environment Agency have put on their website, the closing submissions so that they're available, particularly for those who are interested in this particular case. And it was a big question as to whether Anglian Water licence would end up having some sort of—for lack of a better expression—collateral damage to it, I suppose.

Interview RSA.REG4

So another very topical one which I'm not sure I should mention but I will. We've got one in the Broads in the Ant Broads and Marshes. So, there is again a water company licence that went through the Review of Consents and was considered acceptable. This is Catfield and then through further information that was provided partly by the landowner, it's then come back into the process and whilst we're currently dealing with the agricultural abstraction that is also associated with it, the EA will need to reconsider that licence. So, yes, there are instances where new data becomes available or the groundwater model didn't pick up the sensitivities of the site.

(…)

There may be some kind of water quality information but generally not. In the example of Catfield, they looked at hydrochemistry and the potential interactions of the groundwater to the surface water and the changes. But I would say generally water quality and water resources were considered separately and then through the Review of Consents work, they considered in combination but not in a detailed way I don't think. So you're not putting all of that information into one document.

Water Company

Interview RSA.WC2

That's a very live issue at the moment; whether you want to use that one as an example would be interesting. A little potted history for you—we've got an abstraction point up there at a place called Ludham that we've again been looking at probably in AMP3 originally, so 2005, that was 3 I think, to look at what the impact of that abstraction had on … it's one of the Broads down there, I forget the name of it now, but anyway there's a Broad nearby, and that went through the Habitats Directive process and it basically came out with a clean bill of health.

There was some debate around it but it was generally deemed okay. The agricultural licences were also in the mix at that stage and again it was generally deemed acceptable as to where we were. We agreed to do some work in AMP4 so it did drive some investment. We reduced our abstraction licence slightly and we brought some more water up I think, again from our Norwich zone or one of those zones, so that was generally deemed acceptable.

… There was another piece of Broad and it came to light very late in proceedings, that was considered to be … and this was raised by the Broads Authority that they were concerned about and that hadn't been addressed sufficiently. An awful lot of debate and discussion and certainly I know it's exercised a lot of EA minds over the last year or so on this.

We haven't particularly been involved in those discussions because what's triggered it was the renewal of one of the agricultural licences, and it goes through a period of consultation and advertising and it was that licence application that was challenged. What is happening now is that we are being drawn into that, there's an in-combination impact and we are part of the in-combination impact but our licence as yet to date nothing formal has come of that.

We've had some letters and we've reached an agreement with the EA to say that as an interim measure, you haven't flagged this soon enough for us to put a solution in place now, we will do something in AMP7 and as AMP6 we will reach an interim solution that we try and reduce our abstraction on Ludham, recognising that you've had these challenges from elsewhere. It's a very interesting case study. I'm concerned about where it's taking us, because it is fundamentally questioning what's happening in terms of all this groundwater modelling and where we've got to previously, and that's my concern and I think the EA's concern on that as well.

So you find yourself in some silly situations sometimes where it might be the original boreholes that maybe ... not in this case, Ludham they're all quite local, but there are certainly some cases where it's the original boreholes which aren't necessarily being challenged through the renewal, the variation process.

Anyway we as a company wouldn't say, no, we'll just carry on abstracting from those boreholes because in theory it's a licence of right, we can. Again we wouldn't do that, but you do get yourself into some silly situations. It's just the quirk I suppose of the licencing system. I think that's where we are at Ludham, I think there is that variation piece that we would be renewing at some stage but it is a little way off. What we've essentially said up there is that, fine, we recognise ...

It's very difficult when you get into these sorts of challenges, and you go back to ... there's quite a few high profile and high powered people involved, and various experts been employed here and here on each side and you get two people arguing, it's quite easy, as I'm sure you appreciate, to pick holes in an argument, in a model and things like that. It doesn't take long before you can start unravelling a model and say that's where I'm concerned. It's the best tool that we have available now, do we really want to be unpicking these things? I'm not comfortable with where we could end up, but I know that's where the challenges are coming in at the moment.

Abstractor Groups

Interview RSA.ABS2
Catfield is a farm with two licences that were first granted in 1985 and in 2010 and questions were raised about whether they were having an unac-

ceptable impact on the neighbouring fen, Catfield Fen. One of the many interesting things for me is this concern about just how complicated all this stuff is going to get in the future because what I realised, was that this highly sophisticated Environment Agency groundwater model that I've already mentioned had investigated these two Catfield licences and decided that it wasn't having an impact on the fen. I am pretty sure the Agency might reject this but I think that the model still shows that.

There is a lot more to Catfield than meets the eye. So the Agency was minded to, after five years of investigation, this time last year or some time in 2014, was minded to grant the licences and then new evidence came forward and, to my mind, the evidence revolved around ... the opposition to grant the licence questioned the Agency's groundwater model and it questioned it on two counts. Firstly, that the model is only capable of looking at the environmental impact of the cumulative effect of licences in a catchment, and it is just not possible to run this type of model to understand the impact of these one or two licences. So that's introducing an element of doubt. That was the first one.

Then the second one revolved around the environmental impact being as a consequence of water quality, namely water pH, and the question mark around where the model that's all geared up to understand the relationship between abstraction and flow, whether it was capable of understanding water quality. You'll appreciate that I'm talking to you as a layman, but I think that when those two questions were introduced by those who opposed the granting of the licence, that took us into the realms of precautionary principle, because then the Environment Agency says we cannot show sufficiently that the abstraction isn't having an adverse effect, therefore we no longer feel able to grant the licence. As I say, that is my layman's explanation on what's happened.

The water quality thing I think is really very interesting because the way it's been described to me is that Catfield Fen is obviously peat because it's a fen and it's sat over a saucer of clay, and below the clay there's a layer of chalk aquifer. I do not think it is disputed that there are holes in the chalk that hydrologically link the peat with the chalk. It is just a question of how many there are and where are they, and you'll never know that. So, if we have the farm and then the fen, and the farm is drawing up water from a borehole and that is presumably drawing water across the farm from under the fen. As I say, I still think the groundwater model is saying that that is not having a significant impact in terms of water volume. There is an enormous Anglian Water borehole here. So the issue about Catfield is it is having an in combination effect. If Anglian Water did not exist, presumably the farm would be okay. But this is a licence of right here so the Agency, unless it gets its act together and does something under the RSA programme...

So what happened here, what happened on the fen was that conservation organisations and the owner of Catfield Fen raised concerns about the Fen Orchid which needs an alkaline condition in which to grow. So the supposition is that the abstraction is pulling away alkaline groundwater and that the fen is benefitting from less alkaline water but more acidic rainfall. So that impact is changing the pH of the soil on the fen which is having an adverse impact on the population of Fen Orchid, and the precautionary principle means that it's the responsibility of the farmer to prove that that's not happening.

The farmer is reserving the right to go to public enquiry but the licence has been refused he farmer is talking to the Environment Agency about whether he can construct a reservoir but that is not looking particularly possible. So he's going to need to think about gradually moving out of irrigated crops, which has income, employment and asset value implications. So if this doesn't happen, his farm is considerably devalued. But as long as the legal process continues he is still allowed to irrigate.

Interview RSA.ABS3

There's a site at Catfield Fen, they're all the same. It is individual farmers and collectively we do not have the resources or the knowledge to fight anything on biological grounds and you'd have to pour huge sums of money into it to get anywhere. So essentially in a way we have accepted the biological argument because we can't do anything about it or we think we can't, and everything is done on the mathematical argument of how do we mitigate this to make it as easy for abstractors as possible? Because that is the easier argument to have and that's why you have it.

You can imagine going out to count the number of snails or something that was in the river when the river was high or low, it would be a complete nightmare and you have to take this... you might have to go to an appeal on this and you'd get shot down, you wouldn't have a prayer. Any decent lawyer would just shoot you out the water because you don't have the level of expertise and the level of knowledge because it's always up to the abstractor to disprove what you're being told. That is right, you have got to disprove it. It's not a question of they have to prove, you have to disprove.

Interview RSA.ABS1

A possible cause of the revocation of those licences in Norfolk was the fact that the guy who wanted them revoked was able to finance as many different reports and researchers and solicitors and lawyers as he could care to do, because he's always got an unlimited pit of money where that particular subject was concerned. So he could do that but the farmers, who are on notice really, were unable to provide evidence of their own to refute what's been put forward because they just hadn't got the resources to do it, or to fight the case.

References

Anderson, E. P., Jackson, S., Tharme, R. E., et al. (2019). Understanding Rivers and Their Social Relations: A Critical Step to Advance Environmental Water Management. *WIREs Water, 6*, e1381. https://doi.org/10.1002/wat2.1381.

BAWAG. (2018). Catfield Fen—BAWAG. Retrieved January 8, 2018, from http://www.bawag.co.uk/catfield-fen.aspx.

Bryman, A. (2012). *Social Research Methods*. New York: Oxford University Press.

Butterfly Conservation. (2018). Butterfly Conservation—Catfield Fen, Norfolk. Retrieved January 8, 2018, from https://butterfly-conservation.org/2401-1898/catfield-fen-norfolk.html.

Case, P. (2016). Farmer Loses Fight for Abstraction Licence. *Farmers Weekly*. Retrieved January 8, 2018, from http://www.fwi.co.uk/news/farmer-loses-fight-for-abstraction-licence.htm.

Cash, D. W., Clark, W. C., Alcock, F., et al. (2003). Knowledge Systems for Sustainable Development. *PNAS, 100*, 8086–8091. https://doi.org/10.1073/pnas.1231332100.

Charles, A., Loucks, L., Berkes, F., & Armitage, D. (2020). Community Science: A Typology and Its Implications for Governance of Social-Ecological Systems. *Environmental Science & Policy, 106*, 77–86. https://doi.org/10.1016/j.envsci.2020.01.019.

Earth System Governance Project. (2018). *Earth System Governance. Science and Implementation Plan of the Earth System Governance Project*. Utrecht.

Environment Agency. (2013). Environmental Flow Indicator. What It Is and What It Does. Retrieved September 22, 2017, from http://webarchive.nationalarchives.gov.uk/20140328104910/http://cdn.environment-agency.gov.uk/LIT_7935_811630.pdf.

Environment Agency. (2015). Catfield Fen: Decision on Licence Applications—GOV.UK. Retrieved January 8, 2018, from https://www.gov.uk/government/publications/catfield-fen-decision-on-licence-application/catfield-fen-decision-on-licence-applications#conclusion-and-recommendation.

Fisher, E., Lange, B., & Scotford, E. (2019). *Environmental Law. Text, Cases and Materials* (2nd ed.). Oxford: Oxford University Press.

Grecksch, K., & Lange, B. (2018). *Governance of Water Scarcity and Droughts*. Oxford: Centre for Socio-Legal Studies.

Gupta, J., Akhmouch, A., Cosgrove, W., et al. (2013). Policymakers' Reflections on Water Governance Issues. *Ecology and Society, 18*. https://doi.org/10.5751/es-05086-180135.

Hukkinen, J. I. (2020). Knowing When Knowledge Performs Its Power in Ecological Economics. *Ecological Economics, 169*, 106565. https://doi.org/10.1016/j.ecolecon.2019.106565.

Hulme, M., Lidskog, R., White, J. M., Standring, A. (2020). Social Scientific Knowledge in Times of Crisis: What Climate Change Can Learn from Coronavirus (and Vice Versa). *WIREs Climate Change*, e656. https://doi.org/10.1002/wcc.656.

Ingram, H. (2013). No Universal Remedies: Design for Contexts. *Null*, 38, 6–11. https://doi.org/10.1080/02508060.2012.739076.

ISSC, IDS, UNESCO. (2016). *World Social Science Report 2016, Challenging Inequalities: Pathways to a Just World*. Paris: UNESCO.

Jacobs, K., Lebel, L., Buizer, J., et al. (2016). Linking Knowledge with Action in the Pursuit of Sustainable Water-Resources Management. *PNAS, 113*, 4591–4596. https://doi.org/10.1073/pnas.0813125107.

Jasanoff, S., & Wynne, B. (1998). Science and Decisionmaking. In S. Rayner & E. L. Malone (Eds.), *Human Choice and Climate Change: The Societal Framework* (pp. 1–87). Columbus, OH: Batelle Press.

Kelly, A. (2014). *Catfield Fen Water Abstraction*. Norwich: Broads Authority.

Knill, C., & Liefferink, D. (2007). *Environmental Politics in the European Union*. Manchester: Manchester University Press.

Lange, B., & Cook, C. (2015). Mapping a Developing Governance Space: Managing Drought in the UK. *Current Legal Problems, 68*, 1–38. https://doi.org/10.1093/clp/cuv014.

Lave, R. (2015). The Future of Environmental Expertise. *Null, 105*, 244–252. https://doi.org/10.1080/00045608.2014.988099.

Marsh, T., Cole, G., & Wilby, R. (2007). Major Droughts in England and Wales, 1800–2006. *Weather, 62*, 87–93. https://doi.org/10.1002/wea.67.

Met Office. (2012). Dry Weather during 2003. *Met Office*. Retrieved July 28, 2017, from http://www.metoffice.gov.uk/climate/uk/interesting/2003dryspell.html.

Met Office. (2013). England and Wales Drought 2010 to 2012. *Met Office*. Retrieved July 28, 2017, from http://www.metoffice.gov.uk/climate/uk/interesting/2012-drought.

Pahl-Wostl, C., Knieper, C., Lukat, E., et al. (2020). Enhancing the Capacity of Water Governance to Deal with Complex Management Challenges: A Framework of Analysis. *Environmental Science & Policy, 107*, 23–35. https://doi.org/10.1016/j.envsci.2020.02.011.

Tengö, M., Brondizio, E. S., Elmqvist, T., et al. (2014). Connecting Diverse Knowledge Systems for Enhanced Ecosystem Governance: The Multiple Evidence Base Approach. *Ambio, 43*, 579–591. https://doi.org/10.1007/s13280-014-0501-3.

Williams, C., Fenton, A., & Huq, S. (2015). Knowledge and Adaptive Capacity. *Nature Climate Change, 5*, 82–83. https://doi.org/10.1038/nclimate2476.

Yin, R. K. (2009). *Case Study Research: Design and Methods*. Los Angeles: Sage.

PART III

Outreach

CHAPTER 5

Public Engagement with Drought and Water Scarcity Research

Abstract This chapter makes a contribution to the discussion on public engagement with research. Hence, the focus is less on the research topic as such but more on the communication and dissemination of drought and water scarcity related research. This chapter presents public engagement with research activities: two water-related walks. First, a waterways walk through Birmingham, UK, that stopped at various places and discussed drought and water scarcity issues with participants. Second, a drought walk through St James's Park in London used to disseminate a primer document and highlight issues of public water supply and its function in public spaces. This chapter presents these activities to frame a necessary discussion about the need for better communication of drought and water scarcity in the UK against the background that the issue is often neglected in public debates about water. At the same time water resources managers at the UK's water companies report difficulties in communicating the issue despite having supporting data from monitoring and modelling. Hence, a public drought and water scarcity discussion is overdue and, as the examples show, can come in various shapes.

Keywords Drought • Water scarcity • Public engagement • UK • Waterways walk

© The Author(s), under exclusive license to Springer Nature   93
Switzerland AG 2021
K. Grecksch, *Drought and Water Scarcity in the UK*, Global
Challenges in Water Governance,
https://doi.org/10.1007/978-3-030-65578-5_5

INTRODUCTION

The UK has a drought and water scarcity communication problem. One could say that while floods stay in the head, droughts are easily forgotten. Although the UK has experienced several droughts in the past 20 years as mentioned in the previous chapters, it is the drought of 1976 that people seem to remember. And, people in the UK tend to demonstrate feelings of nostalgia towards past, hot summers where positive feelings towards these events can foster perceptions that people feel safer than they really are (Howarth et al. 2019). Moreover, the experience of drought and water restrictions in the past affects the response to restrictions in the future (Manouseli et al. 2018). However, a drought 45 years ago is not in the memory of those born after 1976 or even those born in the years shortly before 1976. Yet, with drought and water scarcity being an issue that, according to the UK Climate Change Risk Assessment, needs more action (Committee on Climate Change Risk Assessment 2016), this should involve not only technological supply and demand side measures but also the communication of drought and water scarcity. Just like floods, it needs to stay in peoples' heads that this is a serious issue in the UK exacerbated by climate change, urbanisation and population growth. The previous chapter already highlighted the case for a better integration of local knowledge into the generation of drought and water scarcity related knowledge and decision-making. In this chapter the aim is to demonstrate the value of public engagement with research, that is informing, disseminating, communicating and discussing research results with different publics. Public engagement goes beyond simply disseminating research results in the hope that it changes behaviour by building relationships with publics (Cook and Overpeck 2019). Public engagement with research must not necessarily mean the engagement with the general public but also with specialist publics such as those working in the water sector. This will be the case presented here. This chapter, after introducing the concept of public engagement, will present two public engagement with research activities: first, a waterways walk through Birmingham stopping at various water-related places to initiate a discussion about drought and water scarcity; second, a drought walk through St James's Park in London to disseminate the results of a research project to stakeholders and to have a discussion about the role of public parks in water education. In both cases the hypothesis was that talking about water issues at places that are related to water, that is a canal, river, fountains, lake, and so on, inspires a more fruitful discussion because being at these places may conjure up memories, thoughts and reflections that would otherwise not come to mind in a discussion held in a conference room or office. For example, in ethnography artefacts are used to conjure up memories.

Abbott et al. (2019) discuss the importance of water education using the example of the water cycle. This well-known tool comes with several flaws and could, according to the authors, 'undermine efforts to promote an understanding of water and also of general scientific thinking' (ibid., p. 533). The authors especially criticise that 85% of the analysed water cycle diagrams showed no interaction between humans and the water cycle despite the fact that humans dominate critical aspects of the hydrosphere and 80% of the world's population facing water insecurity or severe water scarcity (ibid.). They conclude: 'The omission of humans and associated changes from water cycle diagrams is deeply problematic because it implies that one of our most essential and threatened resources is not influenced by our actions' (ibid. p. 539). The United Nations University's Institute for the Advanced Study of Sustainability suggests the concept of water literacy. Thereby they refer to 'appropriate knowledge about various aspects of water use and management in order to ensure safer water consumption and to contribute to Disaster Risk Reduction (DRR)' (Yuto et al. 2014). Water literacy can be acquired by obtaining basic literacy competencies and a certain level of education, and they suggest, for example, to improve the water literacy of all local governments and residents through educational activities to ensure safe water use and to promote sustainable water supply. Robins et al. (2017) also want to create a more water-literate society, where UK citizens better risk and engage in decision-making about how water should be managed. 'A more water-literate society will better enable water managers to shift from reactionary, crisis-driven approaches to long-term, agenda-driven plans in line with agreed strategic goals' (Robins et al. 2017, p. 52).

There is work and experience available regarding floods and flood resilience in the UK. McEwen et al. (2016) speak of 'watery senses of place', meaning living with water and water issues such as flooding: 'Clearly, post-flood learning needs incorporating into community flood education to increase adaptive capability' (ibid., p. 15). The authors introduce the concept of 'flood memorialisation', described as the process by which 'facts' of the event such as high water levels are recorded and the (emotional) memory of flood impacts is honoured (ibid., p. 19; emphases in the original). The same could be introduced for post-drought learning, and the two walks discussed in this chapter could serve as an example of that. With reference to the second walk, droughts could be 'memorialised' in public parks, serving as a reminder of drought and water scarcity. Also with reference to flooding, Whatmore (2009) (but also Landström et al. [2011]) introduced so-called Environmental Competency Groups (ECGs). ECGs encourage scientists and local residents to work together to create knowledge about local environmental issues. The approach therefore creates a space where those who are directly affected can question expert knowledge and bring their experiences to bear on how the problem is framed and what different courses of action are available (Grecksch and Landström 2021). Again, the two drought walks detailed below offer a possibility to do this, that is listen to expert knowledge but also question it as well as providing a space for bringing in ideas and knowledge about drought and water scarcity.

There are, of course, ample educational and water efficiency programmes across the UK carried out by water suppliers (Grecksch and Lange 2019). However, the focus is on technological devices (e.g. water meters, tap aerators and hippo bags), financial incentives and educational programmes for school children. There is less focus on drought though. Communication about drought is usually occurring when it is already happening and also occurs within the legal framework. This means that a Temporary Use Ban (TUB), that is the prohibition of certain water uses such as washing a car or watering the garden, has to be advertised in newspapers, radio and television ahead of its implementation (Lange and Cook 2015). There is no continuous conversation about drought and water scarcity though in the UK, which could raise awareness and better prepare people for actual drought events and water scarcity situations.

As mentioned before, the aim of this chapter is to raise awareness for public engagement with research, in this case for drought and water scarcity. A secondary aim is to demonstrate how important it is to involve water professionals, that is people working in water companies, water consultancies or regulatory bodies, in these exercises as they are the intermediaries and transmission belts between those who professionally care about water and citizens. Both walks, which will be discussed in this chapter, were primarily aimed at water stakeholders, that is the water professionals just mentioned. Furthermore, both walks were part of conferences aimed at researchers as well as professional stakeholders, and in this sense, they could be easily recruited for the walks. The two walks described and discussed in this chapter are only the beginning of a much-needed public engagement with drought and water scarcity in the UK and hence provide only a snapshot instead of a representative picture of the UK. However, the participants of both walks came not only from different organisational backgrounds but also from different geographical regions of the UK. The next section briefly introduces the concept of public engagement with research followed by a detailed description and analysis of both walks.

Public Engagement: Role and Significance

Public engagement with research describes the many ways that members of the public can be involved in the design, conduct and dissemination of research (Chikoore et al. 2016 provide a history of public engagement in the UK). It is meant to inform, listen, consult, inspire and collaborate with the public. UKRI, the UK's public body bringing together the UK's seven research councils, describes public engagement similarly as: 'Creating opportunities for people to discuss, create and participate in research and innovation is an important way to achieve this, because this makes research and innovation more relevant, impactful and trusted' (UKRI 2020). This means it is a two-way process. It is not simply researchers communicating research results to the public, but it is about a meaningful discussion about research. This can create legitimacy and trust, and it could also inspire new research. Public engagement with research is now an integral part of universities, and specialist teams offer support and training in public engagement with research. Especially for early career researchers, public engagement with research is seen a key transferable skill enhancing their future employability. There are conferences specialising on public engagement, and there are prizes and awards for successful public engagement.

The possibilities for public engagement with research seem endless. Chikoore et al. (2016, p. 160) carried out a survey on public engagement activities among UK academics and ranked them accordingly. The top three are 'presenting to a professional audience', 'presenting a public lecture' and 'writing for a non-academic publication'. However, there are countless more creative engagement activities such as podcasts, blogs, using social media, working with schools, web-documentaries, being interviewed for TV or radio, public performances, exhibitions, science slams or 'pint of science' events. Of course, the type of engagement activity depends on the discipline and research, and not every type of engagement activity is suitable for every type of research. In the cases presented here I decided for walks to discuss and present research and engage with the public (see below).

'The 'public' are all non-academic audiences (including the general public) that can potentially be engaged with by academics' (Chikoore et al. 2016, p. 148). Hence, there is not 'one' public, but people can be members of multiple 'publics'. For example, a water resources manager working for one of the private water supply companies in England or Wales is part of the 'public' consisting of professionals in the water sector, but they are also a part of the wider public using, for instance, water at home. As researchers we can make use of targeting these different 'publics'.

DROUGHT WALKS

The idea to use walks to discuss research or to gather data is not entirely new. Ambrose (2020) provides a literature review in her essay on energy walks pointing out walking interviews and the importance of oral histories in this context. Although her walk focused on energy, there is a striking parallel to water when she says: 'Citizens therefore lack the basic knowledge required to make an assessment of the ethical, environmental and economic implications of the choices made on our behalves and our contemporary relationship with energy is arguably one characterised by complete dependency and almost complete ignorance' (ibid.). 'Energy' can easily be exchanged for 'water' in this sentence without losing any of its meaning.

The original ideas for the two water-related walks presented here are somewhat more mundane. First, it was the desire of the author to engage more in public engagement with research and, second, the ambition to take stakeholders to places where drought and water scarcity happened or places where water plays an important role either directly or indirectly. A third reason was to escape the format of presenting research in the usual setting of a conference room and take other researchers and stakeholders on a post-lunch walk. Both walks were part of drought and water scarcity conferences, and the conferences provided the framework for the walks and facilitated the recruitment of participants. The first drought walk in Birmingham was also filmed by a professional film crew and an interview was made after the walk with the author. Footage from the walk and the interview are featured in a five-minute video about the conference that is accessible to the general public. In this case one public engagement activity led to a second one. Both walks also had a positive side effect in that participants had a chance to talk to each other while walking from one stop to another. They could introduce themselves and meet new colleagues in a way that a traditional format like a presentation would not allow.

Moreover, the context of both drought walks was a work task within the ENDOWS project on 'Innovation, communities and corporate water' led by the author. The aim of the work task was to analyse water efficiency in the public sector, especially the role of social norms. The final outcome is a primer document on water efficiency in the public sector aimed at stakeholders in the water sector, regulators and public sector organisations (Grecksch and Lange 2019). Because the results of the project are helpful for the understanding of the two water-related walks, they are briefly summarised in the following paragraphs.

Current water efficiency campaigns and strategies in England and Wales focus on individual households and private businesses. Water efficiency in public sector and large organisations, with a more 'public' dimension than private individual households, such as workplaces is only discussed in a handful of studies. And, the main tools currently used in England and Wales by water companies are water-saving devices and messages to reduce bills. But water-saving behaviour is influenced not just by individual decisions but by social and psychological drivers such as social norms, values, group behaviour and external factors (culture, family behaviour, infrastructure and regulations). The primer presents findings from academic and grey literature and previous case studies about the potential of water efficiency campaigns to contribute to water saving in the UK within public sector and large organisations—universities, schools, hospitals, council buildings, offices and housing associations. These organisations provide significant untapped potential for water saving by virtue of their size and/ or their nature as public organisations. The focus is on the role of social norms, that is community standards, to promote the uptake and effectiveness of water efficiency campaigns.

The research results suggest that engaging with social norms is a key to devising and implementing successful water efficiency campaigns (Grecksch and Lange 2019). Social norms are value commitments that shape water use behaviour. Social norms have become the tool of choice for today's behavioural policy-makers. The inclusion of a social norm in a message can be a way to encourage citizens to carry out a wide range of socially desirable acts. Social norms serve as cues, and they motivate action by providing information about what is likely to be effective and adaptive (Posner

2002; Larson and Brumand 2014; Lede and Meleady 2019). A good example for a social norm is the message we find in almost every hotel room about the re-use of towels. The primer develops nine key recommendations for a successful water efficiency campaign with the public sector. *Understanding why and how water is valued:* It is important to explore what values customers hold towards water and why or why not they engage in water efficient behaviour. Values are influenced and shaped by society, culture and religious belief systems (Sofoulis 2005; Corral-Verdugo et al. 2008; Hoolohan and Browne 2016; Sharma and Jha 2017; Simpkins 2018). *Narratives and stories:* Telling a story or shaping a narrative is of importance. Simple messages such as 'Save, more water' do not get through to water users. Instead, the bigger story must be told, that is water efficiency should be linked to the wider environmental story, for example the water-energy-food nexus (Waterwise 2018). *Framing:* How information about water efficiency is shaped and contextualised within a familiar frame of reference and meaning is of great importance. Second, it makes a difference who is conveying the message about water saving—water companies, regulators or intermediaries (McQuail 2005; Byerly et al. 2018; Whiting et al. 2019). *Setting realistic targets:* There is a limit to water conservation as we, for example, need to use water to wash ourselves or to wash clothes. People may need water for religious reasons and some people simply do not care about efficient water use (Siero et al. 1996; Steg 2008; Ek and Söderholm 2010; Mills and Schleich 2012). *Competition:* Competitions can be a useful tool in the context of social norms. They can leverage the power of social norms. People like to know where they stand compared to others and they like to be told that they are good (Siero et al. 1996; Petersen et al. 2015; Vine and Jones 2016). *Reference groups:* Our behaviour orientates itself at reference groups—group thinking. In other words, we tend to adapt our behaviour according to what is the norm within a reference group (Goldstein et al. 2008). Herein lies a huge potential for water efficiency campaigns in the public sector. *Align structural and behaviour change measures:* Water-saving devices or new plumbing and the implementation of social norms to change behaviour can go together. In addition, water-saving devices combined with very specific and targeted behaviour change messages work very well together (Goldstein et al. 2008; Steg 2008; Ek and Söderholm 2010; Russell and Fielding 2010; Mondejar-Jimenez et al. 2011; Roccaro et al. 2011; Mills and Schleich 2012). *Building water-saving messages on energy saving campaigns:* A huge factor discouraging people from water-saving behaviour is the fact that they have less control over water infrastructure as, for example, compared to energy. However, a message that targets shorter showers could focus primarily on the fact that it saves energy (Petersen et al. 2015). *Data and evaluation:* Having a good data basis and regularly evaluating the effects and results of water efficiency campaigns that involve social norms are a precondition for successful water efficiency strategies and campaigns (Orr et al. 2018).

The public sector in the UK employs almost six million people. Hence, there is an opportunity for the public sector to act as a role model for other sectors, such as the private sector, the third sector, and private households. A large majority of the workforce spend their days at workplaces where they use water for washing hands, in the office kitchen, in the canteen and for showering, the latter in particular if there is an increase in cycling to work. Furthermore, there is an opportunity for the public sector to carry out a 'multiplier' function. If water-saving behaviour is implemented at work, this behaviour may also be applied at home, but also vice versa. People who engage privately in water-saving behaviour may have an influence upon their peers in larger organisations in which they may work. Water saving is both a public-private sector task. Public sector organisations are well placed to start water-saving behaviour initiatives themselves, for example, as a competition among departments or in the context of staff engagement weeks or by including water efficient appliances in their procurement activities. There is also scope for water companies and the public sector to increase their cooperation on this issue.

Both walks were carefully planned and organised. The author selected the route and tested it before the actual event to make sure it is feasible and walkable within the given time limit and also to have alternative locations in case of heavy rain. Moreover, a health and safety risk analysis was carried out since both walks included crossing streets with heavy motor traffic and the walk in Birmingham included walking along a canal. In this instance the author led the group along the canal, and the co-convenor followed after the group making sure no one accidentally slips and falls into the water. The author also carried a first aid kit and a mobile phone to contact emergency services should that become necessary. All participants had to register before the event and were given detailed information about the walk, including the distance, a map and to wear appropriate clothes and shoes. Although the chosen locations were set, in the sense that they are not movable, it is important to find places where a small group can gather and talk without being a nuisance to the public and also avoiding to talk loudly or even scream because of surrounding noises.

Waterways Walk Birmingham

The waterways walk took place on 14 March 2018 in Birmingham as part of the ENDOWS Showcase event. The event was an opportunity to engage with research and outputs from the Research Councils UK (RCUK) Drought and Water Scarcity Programme and to help shape the final phase of activity. The activity offered an informal and unusual way of discussing and contributing to the question how water efficiency campaigns can promote public social norms in relation to valuing water. Stakeholders and participants had the possibility to showcase to us how they perceive water efficiency campaigns to address their concerns in relation to the water environment. Since the work task started later in 2018, the walk was also used for generating ideas and to sharpen the research questions of the work task.

The 90-minute walk covered 1.8 miles and made stops at defined points along the route and discussed ideas about water efficiency campaigns. The stops served as 'anchor points' to discuss certain aspects such as follows:

What is the value of water?
How do people experience water efficiency campaigns?
What role should citizens play in the management of drought and water scarcity?
What water efficiency campaigns would you be interested in contributing to?
Have you participated in water efficiency campaigns at your workplace?

Each stop was introduced followed by a discussion. Participants were free to intervene and ask questions at any time. A positive side effect of the walk was, as mentioned before, that participants were also able to talk to each other during the walking; hence, the walks also foster stakeholder relationships. Including the two convenors, ten people participated in the walk, ranging from regulatory bodies, third sector organisations to research councils.

The walk's first stop was at a series of canal locks at the Birmingham and Fazeley Canal, one of the many canals that run through Birmingham (Fig. 5.1). We chose this location to present the oral history of a Yorkshire-based woman and keen boater who recollects her memory of an incident in the winter 1984/1985 where she was frozen in the canal in Tipton and

Fig. 5.1 Canal locks Birmingham (Source: The author)

Wolverhampton for a month each when she was moving her boat from Birmingham because the canal was due to be drained. She remembers how they coped with the situation and how it affected working boats as well. This oral history presented an unusual and unexpected aspect of water scarcity—being frozen in and not being able to navigate the canals. It was important to stress the variety at which water scarcity can affect us, and the cold spell in the UK at the beginning of March 2018 saw pipes burst and water companies distributing water bottles among its customers.

The second and third stops were at Victoria Square in Birmingham's city centre. Against the backdrop of Birmingham's council house and town hall, participants were briefly introduced to Birmingham's water supply history. Birmingham is an interesting case because it was the first city to 'municipalise' water 1876. Then mayor Joseph Chamberlain decided to buy the water works and run it for public profit. The 'Birmingham Corporation Water Department' was responsible for Birmingham's water supply from 1876 until 1974. In 1892, land was purchased in mid-Wales and work began at what was to become Elan Valley Reservoirs and the Elan aqueduct. Both still supply Birmingham with water. Built between 1896 and 1906, consisting of five lakes and a 73-mile-long aqueduct, every day 365 million litres travel to Birmingham by gravity alone. It was built because the average rainfall in mid-Wales is almost three times higher than in Birmingham. Today the reservoirs are managed by Welsh water. Cannon Hill Park in Birmingham hosts a model of the Elan Valley Reservoir, and the model was constructed as a tribute to the pioneers of the scheme and opened in 1998.

We also used this stop to introduce the Consumer Council for Water (CCW) whose offices are located at Victoria Square in Birmingham. The CCW is a public body that represents water and sewage consumers in England and Wales. It provides impartial advice and advocacy, and it takes up unresolved complaints. In a recent publication they stressed the fact that customers also have an important part to play when it comes to saving water and making the water resources more resilient. However, they also attest customers a lack of awareness of the pressures of the water system (Consumer Council for Water 2017). We used both the brief introduction into where Birmingham's water comes from and the CCW's role to discuss about whether we value water, whether it matters where water comes from and also if water efficiency campaigns, which usually address domestic customers, should also address the places where most of us spend our daytime—at our workplaces. In other words, do we need to widen the focus of water efficiency campaigns?

The fourth stop was planned outside Ofwat's offices close to Birmingham's New Street railway station. Instead, the lively discussion at Victory Square was further continued, and Ofwat's role was briefly introduced. Ofwat is a key actor and responsible for economic regulation of water industry in England and Wales. We therefore briefly discussed the role of Ofwat as a catalyst for innovation and whether Ofwat could play a role in animating water companies to take more care of the public sector and water efficiency.

The discussion among the participants during the walk contributed to the discussion about water efficiency campaigns and can be summarised as follows. Water scarcity needs to be put into a wider context. Water scarcity is about human behaviour, technological infrastructure and weather events. It is important to emphasise and talk about water scarcity as it occurs more often than drought. Moreover, there is a challenge of positive incentives for customers to save water, especially against the background of increasing water bills. Furthermore, the question was discussed, where does water come from and is that important? Is there such a thing as 'local' water compared to the discussion about local food. Lastly, the fragmentation of the water governance system and how it affects water efficiency campaigns was a point of discussion. These discussion points were helpful in shaping the research project, which, as mentioned before, was only due to begin later that year. In fact, it sharpened the research question and direction of the research towards a widening of water efficiency campaigns (see summary of the research in the previous section). We did not explicitly measure feedback for this walk as we did for the second walk, but the aforementioned video of the conference, which features the walk, also contains the statement by one of the participants: 'I really enjoyed the canal walk we did' (About Drought 2018 Minute 1:39).

DROUGHT WALK LONDON

The drought walk took place on 7 November 2019 in London as part of the ENDOWS About Drought Download Event. The event marked the end of the ENDOWS project and presented the results of the various work tasks to stakeholders from water companies, regulatory bodies, consultants and researchers. The walk was organised as a lunchtime walk through St James's Park, the closest park to the Royal Society, the conference venue. As opposed to the first walk in Birmingham, this walk was only 45 minutes long. The idea behind walking through St James's Park was to learn about its water features and function and to talk about water efficiency in the workplace. Hence, the walk was used as a dissemination event for the primer on 'Water efficiency in the public sector. The role of social norms'. As explained before, the Primer discusses social norms, such as community standards, as an instrument to instigate more efficient water-saving behaviour in organisations with a public dimension—universities, schools, hospitals, and council buildings. We wanted stakeholders to share their thoughts, ideas and experiences about social norms and water efficiency in large organisations in an informal and energising format.

Out of overall 150 conference participants, 22 joined the walk ranging from representatives of regulatory bodies, water companies, water consultancies, research councils and researcher to PhD students. At the beginning of the walk, participants were briefed about health and safety, the length of the walk and what to expect. As with the first walk, it was hypothesised that visiting places related to water and water efficiency is a great opportunity to develop conversations about water efficiency that may also draw on comparisons between different practices in relation to water saving, including those in various participants' workplaces.

The first stop was the water fountain in St James's Park. The stop was used to introduce the convenors, to have a very quick round of introductions of the participants and to explain the purpose of the walk. As mentioned before the walk was primarily used to disseminate the primer document, but a secondary purpose was to combine this with a discussion about the role and function of public parks and their potential to be used for water education. Thus, this first stop introduced the history of St James's Park. St James's Park is the oldest royal park in London and surrounded by three palaces (Buckingham, St James and Westminster). The park dates back to 1532 when Henry VIII acquired the site as yet another deer park. James I improved the drainage and controlled the water supply by bringing water over from Hyde Park. He also kept a collection of animals in the park among which were camels and crocodiles. But it was Charles II (1630–1685) who made dramatic changes to the park, including a complete redesign of the lawns, trees and avenues. This was inspired by elaborate French gardens. The park became more formal. The centrepiece was a 2500-ft-long canal lined by trees. He also opened the park to the public. During the Hanoverian period, parts of the long canal where filled creating what today is Horse Guards Parade. The next complete overhaul and redesign of the park was undertaken by John Nash in a more romantic style, that is the canal was transformed into a natural-looking lake. This is pretty much how the park still looks today.

We then combined this history of the park with the results of our research and communicating water issue in the UK in general. For instance, one of the challenges we found during the research about drought and communication about water resources is the limited visibility of the issue of periodic water scarcity in the UK. Hence, parks are natural places for starting conversations about drought and valuing water. Public parks with water features provide an opportunity to render more visible the issue of water availability for a range of citizens also in urban areas, that is by showing changes in water levels of the water features and vegetation, for example during periods of drought. During the national COVID-19 lockdown, the role of public parks and access to them was further highlighted; hence, there are possibilities of emphasising the recreational function of parks and its potential role in water education. Already media images of hot and water scarce summer periods usually show citizens congregating in parks and on beaches in the UK, with children playing in water fountains.

The second stop was at the blue bridge that crosses the pond in St James's Park. Here we discussed first the role and function of public parks. Public parks fulfil a vital function in cities, towns, and also villages (village green). They are places to relax, to play, to rest and to eat. Public parks were created for the population's well-being. They also provide space for wildlife and biodiversity in urban areas ('lungs for the city'). Water features play a central role in public parks either as lakes, ponds, fountains or on children's playgrounds. Public parks are an important feature of public life, where people, including children, meet, where we can pause and start new conversations. Hence, harnessing a public sphere for promoting water efficiency campaigns was also the starting point for our research on water efficiency in the public sector.

As mentioned in the introduction to this chapter, the focus of water efficiency campaigns in the UK is on domestic households and businesses, leaving untapped potential for engaging with the public sector in water efficiency campaigns. We concluded this stop saying that parks are natural spaces where we can also see and experience some aspects of a changing climate, for example earlier change of colours of trees during times of water scarcity, changing wild life. Hence, parks can be an opportunity to get us to think and learn about the impacts of a changing climate on the management of water resources in the UK.

The third and final stop was at a public drinking water fountain located in the park. This public water fountain was installed by the Metropolitan Drinking Fountain and Cattle Trough Association (Fig. 5.2). The association was set up in London in 1859 to provide free drinking water, and the organisation is still active, though under the name Drinking Fountain Association (The Drinking Fountain Association 2017). Public water fountains and the availability of free drinking water have somewhat come back to the centre of attention in recent years in the context of the discussion about plastic bottles and their impact on the environment (Hawkins 2017; Heathcote 2018; Tosun et al. 2020). For example, the current mayor of London, Sadiq Khan, published plans for a new drinking water fountain network in 2017 (BBC News 2017). We mainly used this stop to briefly introduce the primer and its nine key recommendations (see summary above).

A new element during this walk was to get concrete feedback from participants about the walk. For this purpose, feedback cards were

Fig. 5.2 Public drinking water fountain (Source: The author)

distributed at the beginning of the walk and collected at the end of the walk or later. The feedback cards asked for the type of organisation the participant represents (water company, regulator, research or other), but participants could also choose not to share this information. This was followed by a question on the overall assessment of the walk ranging from 1 (poor) to 5 (excellent). We also asked participants to describe the experience in three words before asking them how useful they found the event in finding out about the primer document. The last question asked how the primer may be relevant for their work. Out of 22 participants, 8 returned the feedback forms. Of those eight, one participant was from a water company, one from a research, five were 'other' (communications, consultancy, PhD student, research council and one graduate student). One participant did not disclose their organisation. Regarding the overall experience, six participants selected '4' and one participant selected '5'. One participant did not answer this question. Asked to describe the experience of the drought walk in three words or phrases, we received the following eight answers: 'informative, refreshing'; 'bright, breezy, informative'; 'refreshing, stimulating, entertaining'; 'interesting, engaging, informative'; 'interesting, unusual, networking'; 'interactive, local, networking'; 'informative, difficult to hear sometimes' and 'excellent, informative, nice'. The question on 'how useful participants found the event in finding out about the primer?' was answered by six participants; the overall answers were positive: two participants responded 'good', 'very/very useful' was mentioned by two participants and other participants replied, 'good, wasn't previously aware of it' and 'useful—I download a copy.' The final question 'How may the primer be relevant to your work?' was answered by seven participants: 'science dissemination'; 'could be a story for the environment (CIWEM)'; 'social norms will be key for changing water use behaviour'; 'something our clients are very keen to explore'; 'just informative'; 'would be good to give to Estates & Facilities in the universities to encourage them to improve water efficiency'; 'not really, but interesting nonetheless' and 'the literature review will be useful for our own on (unreadable word) piece on water'. Thus, the overall feedback for the walk was very positive and encouraging. It was also widely shared on social media on the day, and the positive response is also an encouragement to organise similar events in the future. The majority of respondents evaluated the walk and using it to disseminate research results as 'very good'. It was also interesting to see that participants mentioned 'networking' as an item when they described the walk. As mentioned before, this is a positive side effect of the walks, as stakeholders have the opportunity to introduce themselves and talk to each other while walking.

Self-critically it must be said that the time for the walk, 45 minutes, was very short and we were not able to get through all we had planned. For example, we had prepared questions for the participants: Do you have any experience with water efficiency at your workplace? Who in your organisation is responsible for water efficiency? And would that person be interested in the primer? Which policy could this issue be tacked on? We were hoping for a fruitful discussion especially since the walk was very popular with conference participants. The second challenge we faced was the noise. St James's Park is popular with tourists on their way to or from Buckingham Palace. Speaking to the walk participants was sometimes interrupted by large tourist groups speaking or crossing our paths. However, the feedback we received was nonetheless positive and encouraging to repeat these events.

Conclusion

This chapter presented two drought and water scarcity related walks as a tool for public engagement with research. Public engagement with research offers the opportunity to engage with different publics about research and its results. It is a two-way process and can also be a useful tool to gather ideas before the start of a research project as the first walk has shown. Both walks were successful, and the feedback from participants was very positive. Walks that discuss research with the public or stakeholders need to be carefully planned and facilitated. In both cases it helped that the walks were part of conference activities, which is helpful for recruiting participants. They also require flexibility. The weather can change, a street is closed or time is running out. As convenor one must be able to adapt quickly. In this regard the second walk was, as explained, challenging. Limited time was a constraint and some discussion points had to be dropped. Nonetheless, walks are highly recommendable as they offer the combination of disseminating and discussing research, networking opportunities for participants and the sheer fact of being outside and the place that the research is about.

REFERENCES

Abbott, B. W., Bishop, K., Zarnetske, J. P., et al. (2019). Human Domination of the Global Water Cycle Absent from Depictions and Perceptions. *Nature Geoscience, 12,* 533–540. https://doi.org/10.1038/s41561-019-0374-y.

About Drought. (2018). About Drought Showcase 2018 Highlights—YouTube. Retrieved August 2, 2020, from https://www.youtube.com/watch?v=HYDJGHhy8do.

Ambrose, A. (2020). Walking with Energy: Challenging Energy Invisibility and Connecting Citizens with Energy Futures through Participatory Research. *Futures, 117,* 102528. https://doi.org/10.1016/j.futures.2020.102528.

BBC News. (2017). Drinking Fountains Planned for London. *BBC News.*

Byerly, H., Balmford, A., Ferraro, P. J., et al. (2018). Nudging Pro-environmental Behavior: Evidence and Opportunities. *Frontiers in Ecology and the Environment, 16,* 159–168. https://doi.org/10.1002/fee.1777.

Chikoore, L., Probets, S., Fry, J., & Creaser, C. (2016). How Are UK Academics Engaging the Public with Their Research? A Cross-Disciplinary Perspective. *Higher Education Quarterly, 70,* 145–169. https://doi.org/10.1111/hequ.12088.

Committee on Climate Change Risk Assessment. (2016). *UK Climate Change Risk Assessment 2017.* Synthesis Report: Priorities for the Next Five Years. London.

Consumer Council for Water. (2017). *Water, Water Everywhere? Delivering a Resilient Water System (2016–17).* Birmingham: Consumer Council for Water.

Cook, B. R., & Overpeck, J. T. (2019). Relationship-Building between Climate Scientists and Publics as an Alternative to Information Transfer. *WIREs Climate Change, 10,* e570. https://doi.org/10.1002/wcc.570.

Corral-Verdugo, V., Carrus, G., Bonnes, M., et al. (2008). Environmental Beliefs and Endorsement of Sustainable Development Principles in Water Conservation: Toward a New Human Interdependence Paradigm Scale. *Environment and Behavior, 40,* 703–725. https://doi.org/10.1177/0013916507308786.

Ek, K., & Söderholm, P. (2010). The Devil Is in the Details: Household Electricity Saving Behavior and the Role of Information. *Energy Policy, 38,* 1578–1587. https://doi.org/10.1016/j.enpol.2009.11.041.

Goldstein, N. J., Cialdini, R. B., & Griskevicius, V. (2008). A Room with a Viewpoint: Using Social Norms to Motivate Environmental Conservation in Hotels. *Journal of Consumer Research, 35,* 472–482. https://doi.org/10.1086/586910.

Grecksch, K., & Landström, C. (2021). Drought and Water Scarcity Management Policy in England & Wales—Current Failings and the Potential of Civic Innovation. *Frontiers in Environmental Science.*

Grecksch, K., & Lange, B. (2019). *Water Efficiency in the Public Sector. The Role of Social Norms. A Primer*. Oxford: Centre for Socio-Legal Studies, University of Oxford.

Hawkins, G. (2017). The Impacts of Bottled Water: An Analysis of Bottled Water Markets and Their Interactions with Tap Water Provision. *WIREs Water, 4*, n/a-n/a. https://doi.org/10.1002/wat2.1203.

Heathcote, E. (2018). Drinking Fountains Quench a Thirst for Sustainability. *Financial Times*. Retrieved June 26, 2018, from https://www.ft.com/content/085f675e-6a42-11e8-aee1-39f3459514fd.

Hoolohan, C., & Browne, A. L. (2016). Reframing Water Efficiency: Determining Collective Approaches to Change Water Use in the Home. *British Journal of Environment & Climate Change, 6*, 179–191.

Howarth, C., Kantenbacher, J., Guida, K., et al. (2019). Improving Resilience to Hot Weather in the UK: The Role of Communication, Behaviour and Social Insights in Policy Interventions. *Environmental Science & Policy, 94*, 258–261. https://doi.org/10.1016/j.envsci.2019.01.008.

Landström, C., Whatmore, S. J., Lane, S.N., et al. (2011). Coproducing Flood Risk Knowledge: Redistributing Expertise in Critical 'Participatory Modelling': Environment and Planning A. https://doi.org/10.1068/a43482.

Lange, B., & Cook, C. (2015). Mapping a Developing Governance Space: Managing Drought in the UK. *Current Legal Problems, 68*, 1–38. https://doi.org/10.1093/clp/cuv014.

Larson, K., & Brumand, J. (2014). Paradoxes in Landscape Management and Water Conservation: Examining Neighborhood Norms and Institutional Forces. *Cities and the Environment (CATE), 7*(1), Article 6.

Lede, E., & Meleady, R. (2019). Applying Social Influence Insights to Encourage Climate Resilient Domestic Water Behavior: Bridging the Theory-Practice Gap. *Wiley Interdisciplinary Reviews: Climate Change, 10*, e562. https://doi.org/10.1002/wcc.562.

Manouseli, D., Anderson, B., & Nagarajan, M. (2018). Domestic Water Demand During Droughts in Temperate Climates: Synthesising Evidence for an Integrated Framework. *Water Resources Management, 32*, 433–447. https://doi.org/10.1007/s11269-017-1818-z.

McEwen, L., Garde-Hansen, J., Holmes, A., et al. (2016). Sustainable Flood Memories, Lay Knowledges and the Development of Community Resilience to Future Flood Risk. *Transactions of the Institute of British Geographers, 42*, 14–28. https://doi.org/10.1111/tran.12149.

McQuail, D. (2005). *Mass Communication Theory*. London: Sage.

Mills, B., & Schleich, J. (2012). Residential Energy-Efficient Technology Adoption, Energy Conservation, Knowledge, and Attitudes: An Analysis of European Countries. *Energy Policy, 49*, 616–628. https://doi.org/10.1016/j.enpol.2012.07.008.

Mondejar-Jimenez, J. A., Cordente-Rodriguez, M., Meseguer-Santamaria, M. L., & Gazquez-Abad, J. C. (2011). Environmental Behavior and Water Saving in Spanish Housing. *International Journal of Environmental Research, 5*, 1–10.

Orr, P., Papadopoulou, L., & Twigger-Ross, C. (2018). *Water Efficiency and Behaviour Change Rapid Evidence Assessment (REA) Final Report WT1562, Project 8.* London: Defra.

Petersen, J. E., Frantz, C. M., Shammin, M. R., et al. (2015). Electricity and Water Conservation on College and University Campuses in Response to National Competitions among Dormitories: Quantifying Relationships between Behavior, Conservation Strategies and Psychological Metrics. *PLoS One, 10*, e0144070. https://doi.org/10.1371/journal.pone.0144070.

Posner, E. A. (2002). *Law and Social Norms.* Cambridge, MA; London: Harvard University Press.

Robins, L., Burt, T. P., Bracken, L. J., et al. (2017). Making Water Policy Work in the United Kingdom: A Case Study of Practical Approaches to Strengthening Complex, Multi-Tiered Systems of Water Governance. *Environmental Science & Policy, 71*, 41–55. https://doi.org/10.1016/j.envsci.2017.01.008.

Roccaro, P., Falciglia, P. P., & Vagliasindi, F. G. A. (2011). Effectiveness of Water Saving Devices and Educational Programs in Urban Buildings. *Water Science and Technology, 63*, 1357–1365. https://doi.org/10.2166/wst.2011.190.

Russell, S., & Fielding, K. (2010). Water Demand Management Research: A Psychological Perspective. *Water Resources Research, 46*, W05302. https://doi.org/10.1029/2009WR008408.

Sharma, R., & Jha, M. (2017). Values Influencing Sustainable Consumption Behaviour: Exploring the Contextual Relationship. *Journal of Business Research, 76*, 77–88. https://doi.org/10.1016/j.jbusres.2017.03.010.

Siero, F. W., Bakker, A. B., Dekker, G. B., & Van den burg, M. T. C. (1996). Changing Organizational Energy Consumption Behaviour through Comparative Feedback. *Journal of Environmental Psychology, 16*, 235–246. https://doi.org/10.1006/jevp.1996.0019.

Simpkins, G. (2018). Running Dry. *Nature Climate Change, 8*, 369. https://doi.org/10.1038/s41558-018-0164-3.

Sofoulis, Z. (2005). Big Water, Everyday Water: A Sociotechnical Perspective. *Continuum, 19*, 445–463. https://doi.org/10.1080/10304310500322685.

Steg, L. (2008). Promoting Household Energy Conservation. *Energy Policy, 36*, 4449–4453. https://doi.org/10.1016/j.enpol.2008.09.027.

The Drinking Fountain Association. (2017). The Drinking Fountain Association. Retrieved August 2, 2020, from http://drinkingfountains.org/.

Tosun, J., Scherer, U., Schaub, S., & Horn, H. (2020). Making Europe Go from Bottles to the Tap: Political and Societal Attempts to Induce Behavioral Change. *WIREs Water, 7*, e1435. https://doi.org/10.1002/wat2.1435.

UKRI. (2020). Public Engagement—UK Research and Innovation. Retrieved August 1, 2020, from https://www.ukri.org/public-engagement/.

Vine, E. L., & Jones, C. M. (2016). Competition, Carbon, and Conservation: Assessing the Energy Savings Potential of Energy Efficiency Competitions. *Energy Research & Social Science, 19*, 158–176. https://doi.org/10.1016/j.erss.2016.06.013.

Waterwise. (2018). *Water Efficiency Strategy for the UK. Year 1 Report*. How Is the UK Doing? Waterwise, London.

Whatmore, S. J. (2009). Mapping Knowledge Controversies: Science, Democracy and the Redistribution of Expertise. *Progress in Human Geography; London, 33*, 587–598. https://doi.org/10.1177/0309132509339841.

Whiting, A., Kecinski, M., Li, T., et al. (2019). The Importance of Selecting the Right Messenger: A Framed Field Experiment on Recycled Water Products. *Ecological Economics, 161*, 1–8. https://doi.org/10.1016/j.ecolecon.2019.03.004.

Yuto, K., Eri, Y., Norichika, K., et al. (2014). *Linking Education and Water in the Sustainable Development Goals*. POST2015/UNU-IAS Policy Brief #2. United Nations University Institute for the Advanced Study of Sustainability, Tokyo.

CHAPTER 6

Conclusion

Abstract The conclusion briefly summarises the results of each chapter and fleshes out key lessons and recommendations for drought and water scarcity in the UK and beyond.

Keywords Drought • Water scarcity • UK • Water resources management • Water governance

This book and its chapters provided social sciences perspectives on drought and water scarcity in the UK with special regard to governance, knowledge and outreach. However, the results and conclusions are also relevant beyond the UK context. The focus of the book was on the UK, owing to the focus of the underlying research it is based on. However, the following paragraphs will also reflect beyond the UK context and flesh out how the results could inform the wider debate on drought and water scarcity and water governance in general and beyond the UK context. Although other countries and regions such as Australia, California or South Africa are far more affected by drought and water scarcity, they all require sustainable governance regimes, sound knowledge to inform decision-making and communicating drought and water scarcity to the public in order to be prepared. However, knowledge is content specific and dependent upon what and who is involved in the knowledge production processes

K. Grecksch, *Drought and Water Scarcity in the UK*, Global Challenges in Water Governance,
https://doi.org/10.1007/978-3-030-65578-5_6

(Brugnach and Ingram 2012). And, knowledge processes are specific to particular communities, and what is considered reliable, legitimate knowledge in some communities will have no standing in others (Ingram 2013). Zwarteveen et al. (2017) take the example of the Dutch Delta approach and how it was transferred to other delta countries concluding that it was financial support by the Dutch government for their application elsewhere that was most important. Hence, this opens up the debate about 'which types of knowledge are mobilized in the making of governance decisions, also diversifying the possible performances of tools' (Zwarteveen et al. 2017, p. 7). Therefore, even if knowledge is context specific and dependent, it nonetheless widens the array of options, tools or approaches, and the results presented here should be seen in this context.

The three themes of the book—governance, knowledge and outreach—are key. Drought and water scarcity governance needs to make sure water resources are managed sustainably, enabling all those who are involved in drought and water scarcity governance to take part in decision-making and using a wide variety of drought and water scarcity management options in order to respond to all aspects of drought and water scarcity. For this purpose, Chap. 2 provided an overview over the currently applied drought and water scarcity management options in England and Wales and contrasted it with available options based on a literature and document review. It confirmed that English and Welsh water companies adhere to options that are prescribed by the regulatory framework and options that favour the supply side of water resources management. Overall, it could be concluded that drought and water scarcity options in England and Wales are reactive rather than proactive. This chapter also established a new typology for drought and water scarcity management options that differs from the usual supply and demand dichotomy. Instead, the typology introduced a new classification that covers aspects such as options according to different abstractor groups, overarching frameworks or valuing water. It was argued that this novel typology complements the standard supply and demand dichotomy and could help identifying weak points in current drought and water scarcity management, thereby opening up the opportunity to introduce new options. This could be relevant to other jurisdictions as well. It provides a clear and simple assessment of current drought and water scarcity policy, showing strengths and weaknesses. As such it is heuristic, but water policy-makers could use it as a starting point to assess and improve their policies. These kind of heuristic approaches have been proven successful in areas such as climate change adaptation and water governance before (Gupta et al. 2010; Grecksch 2013).

In order to explore one aspect, the integration of all stakeholders affected by drought and water scarcity, further, Chap. 3 explored how different business and industry sectors in the UK deal with drought and water scarcity. Drought and water scarcity hold severe implications for businesses, industries and their supply chains ranging from slowed-down production to the interruption of production processes over long time periods. The role of drought and water scarcity and the extent to how each sector could be affected unsurprisingly varies across the different sectors. Nonetheless, businesses and industries in the UK are aware of the potential effects of drought and water scarcity because either they experienced drought events before or they are forward thinking in their approach to water resources. The solutions to future challenges of drought and water scarcity are tackled through technological solutions and cooperation with other sectors and regulatory bodies. Especially the latter should be fostered and institutionalised as businesses and industries currently do not belong to the key stakeholders in the UK drought governance space. However, the Water Resources in the South East Group (WRSE), Water Resources East Group in Anglia and most recently the Water Resources North (WReN) group are three organisations in the UK that foster the collaboration between water companies, regulators and other stakeholders in the respective regions. They could serve as a model for other regions as well. So far, how businesses and industries are affected by drought and water scarcity receives little attention in the international literature suggesting that this is a topic that deserves more attention and research also beyond the UK context. Maintaining supply chains, especially in the food and drinks sector, should be a top priority during drought and water scarcity. Businesses and industries have often very specific and long-term knowledge about their water use and could contribute this knowledge in order that water suppliers, no matter if private or public, make better and more informed decisions.

Knowledge, the second theme, was explored in Chap. 4. There is a mutually constitutive relationship between power and knowledge that is crucial in unravelling linkages between the interests that actors pursue and the normative frames and descriptive terms they draw on to represent reality (Zwarteveen et al. 2017). Actors who produce knowledge need to be aware of how knowledge performs its power (Hukkinen 2020). The chapter demonstrated the importance of environmental science knowledge, especially hydroecological data and local (expert) knowledge. Furthermore, the chapter delivered insights into the relationship of the different actors involved in drought and water scarcity management. This relationship is shaped by power relationships and who provides knowledge such as monitoring data and modelling results. Here, the UK and especially England and Wales are unique in the sense that the system of water supply and wastewater treatment is fully privatised. And, the number of actors in regulatory agencies, water companies and consultancies is relatively small. This has implications for the aforementioned power relationships and the production of knowledge. In the case of the UK, the balance of power could, and to some extent already is, easily shift away from the regulators to private water companies and consultancies sped up by the concentrated ownership of UK water companies. And with this shift in power comes a shift in who holds water-related knowledge. Beyond the UK context this should be taken into account or even be a warning to those who promote more privatisation of utilities such as water.

These aspects were discussed using the example of Catfield Fen, a case that showcases the link between the application of regulatory tools and the challenges but also opportunities with regard to what knowledge is applied and which stakeholder provides which type of knowledge. It also shed light on power relationships between key actors in the drought governance space. In this case it showed a knowledge inequality that was shaped by financial resources available to one actor. To avoid knowledge inequalities in water governance, coordinated processes of knowledge production, thereby diversifying the knowledge base and including different voices (Tengö et al. 2014), could lead to agreeable results for all parties involved. A diversification of knowledge and abolishing knowledge hierarchies (Hulme et al. 2020), where, for example, natural science knowledge dominates over social science knowledge, could further reduce knowledge inequalities.

The general desire expressed by stakeholders is to generate and integrate more hydroecological data. Hydroecological data is crucial as monitoring data feeds models and the modelling results inform decisions regarding water demand and supply. But the results also show that data and knowledge can be contested as in the case of the Environmental Flow Index. However, if more data is to be gathered the questions arises by whom? This has implications for the power relationships between actors as well. Hence, will the Environment Agency collect more data or will it delegate this to the water companies, who can also delegate it to specialised consultancies?

A third element discussed in Chap. 4 was the role of local (expert) knowledge. Local (expert) knowledge can fill gaps left by more abstract and formal environmental science knowledges and add new perspectives. And, including local knowledge in the application of regulatory tools for preventing and managing drought and water scarcity can empower stakeholders and strengthen the legitimacy of regulatory decisions. However, so far local knowledge remains a mostly untapped source of knowledge in the UK drought and water scarcity management context. The desire to include more local knowledge in environmental governance issues is of course not limited to the UK (Jacobs et al. 2016; Charles et al. 2020), but issues and questions of participation and access to decision-making processes remain (Grecksch and Klöck 2020).

The third theme, outreach, was addressed in Chap. 5. Outreach activities such as public engagement with research offers the opportunity to engage with different publics about research and its results. It is a two-way process, and in this instance, two drought- and water scarcity-related walks were presented and discussed. Both walks were successful and received very good feedback from participants. Walks are, of course, only one of the sheer endless opportunities of public engagement with research. For the researcher they offer an opportunity to disseminate results in an unusual manner but preferably at places where the research happened or that represent the subject matter. In the two presented cases the aim was to start a discussion at places where drought and water scarcity happened or, in the second case, using water features in a public park to discuss the role and function of public parks in water education. In both cases participants were not from the general public but mainly water stakeholders. However, stakeholders can act as intermediaries between research and the public, though other promising concepts exist as well. Mould et al. (2020) discuss the concept of 'river champions', people who care deeply about a river and are often crucial for the implementation of measures. Often these river champions lead not by power, but they have strong social skills and emotional intelligence. The authors also mention that often, river champions are able to disrupt or unsettle the status quo. Hence, there could be potential for 'drought champions' fulfilling a similar role. Not just in the UK but elsewhere people need to (re)gain a better value of water and water issues. A discussion that should not be influenced and 'forced' by actual drought events such as in Cape Town (Simpkins 2018), but generally a discussion about water. Based on these results the following recommendations can be made:

- Water suppliers/water companies should more actively embrace new options and should also consider going beyond the 'willingness to pay' and 'cost-benefit analysis' horizon in order to tackle future challenges such as climate change, population growth and changing water demand patterns.
- Integrate all stakeholders who are engaged in drought and water scarcity management, not just on an ad-hoc basis and when in drought, but continuously. This could also balance power relationships and avoid knowledge inequalities.
- Pay special attention to stakeholders from businesses and industries, especially the food and drink business. Drought and water scarcity have knock-on effects such as disrupted supply chains that could carry severe implications for public health.
- Listen, integrate and use local (expert) knowledge. It enhances the knowledge base, empowers and legitimises decision-making, and again, reduces knowledge inequalities.
- Start and have a continuous conversation about drought and water scarcity with stakeholders and the public to rebuild trust and the value of water.

I think what has become clear through the research presented here is that drought and water scarcity are issues in the UK. It is a multifaceted issue involving different stakeholders, it requires specialist knowledge and it is an issue that should definitely be communicated more to the general public. Hence, there is ample space and opportunities for improvement as discussed in the chapters and this conclusion. The implications and consequences of the UK leaving the European Union on water policy can only be a matter of speculation at the moment. It remains to be seen whether the UK keeps most of the EU regulations, abolishes them or comes up with forward-thinking ideas regarding the environment. Currently a new Environmental Bill is being discussed, and in summer 2020, the head of the Environment Agency endorsed a proposal that weakens current rules on the cleanliness of rivers based on the EU's Water Framework Directive (Laville 2020). In any case, it is hard to go against the myth of the UK being a 'wet' country where it supposedly always rains. Therefore, the aim should be to integrate discussions about drought, water scarcity, flooding, water in general and the links to other sectors such as energy, agriculture and food production.

REFERENCES

Brugnach, M., & Ingram, H. (2012). Ambiguity: The Challenge of Knowing and Deciding Together. *Environmental Science & Policy, 15*, 60–71. https://doi.org/10.1016/j.envsci.2011.10.005.

Charles, A., Loucks, L., Berkes, F., & Armitage, D. (2020). Community Science: A Typology and Its Implications for Governance of Social-Ecological Systems. *Environmental Science & Policy, 106*, 77–86. https://doi.org/10.1016/j.envsci.2020.01.019.

Grecksch, K. (2013). Adaptive Capacity and Regional Water Governance in North-Western Germany. *Water Policy*, 794–815. https://doi.org/10.2166/wp.2013.124.

Grecksch, K., & Klöck, C. (2020). Access and Allocation in Climate Change Adaptation. *International Environmental Agreements, 20*, 271–286. https://doi.org/10.1007/s10784-020-09477-5.

Gupta, J., Termeer, C., Klostermann, J., et al. (2010). The Adaptive Capacity Wheel: A Method to Assess the Inherent Characteristics of Institutions to Enable the Adaptive Capacity of Society. *Environmental Science & Policy, 13*, 459–471.

Hukkinen, J. I. (2020). Knowing When Knowledge Performs Its Power in Ecological Economics. *Ecological Economics, 169*, 106565. https://doi.org/10.1016/j.ecolecon.2019.106565.

Hulme, M., Lidskog, R., White, J. M., & Standring, A. (2020). Social Scientific Knowledge in Times of Crisis: What Climate Change Can Learn from Coronavirus (and Vice Versa). *WIREs Climate Change* e656. https://doi.org/10.1002/wcc.656.

Ingram, H. (2013). No Universal Remedies: Design for Contexts. *Null, 38*, 6–11. https://doi.org/10.1080/02508060.2012.739076.

Jacobs, K., Lebel, L., Buizer, J., et al. (2016). Linking Knowledge with Action in the Pursuit of Sustainable Water-Resources Management. *PNAS, 113*, 4591–4596. https://doi.org/10.1073/pnas.0813125107.

Laville, S. (2020). Environment Agency Chief Supports Plan to Weaken River Pollution Rules. *The Guardian*. Retrieved August 28, 2020, from http://www.theguardian.com/environment/2020/aug/19/environment-agency-chief-backs-plan-to-water-down-river-cleanliness-rules-james-bevan.

Mould, S., Fryirs, K., Lovett, S., & Howitt, R. (2020). Supporting Champions in River Management. *WIREs Water, 7*, e1445. https://doi.org/10.1002/wat2.1445.

Simpkins, G. (2018). Running Dry. *Nature Climate Change, 8*, 369. https://doi.org/10.1038/s41558-018-0164-3.

Tengö, M., Brondizio, E. S., Elmqvist, T., et al. (2014). Connecting Diverse Knowledge Systems for Enhanced Ecosystem Governance: The Multiple Evidence Base Approach. *Ambio, 43,* 579–591. https://doi.org/10.1007/s13280-014-0501-3.

Zwarteveen, M., Kemerink-Seyoum, J. S., Kooy, M., et al. (2017). Engaging with the Politics of Water Governance. *WIREs Water, 4,* 1–9. https://doi.org/10.1002/wat2.1245.

INDEX

K. Grecksch, *Drought and Water Scarcity in the UK*, Global Challenges in Water Governance, https://doi.org/10.1007/978-3-030-65578-5